WRITINGS
ON
ARCHITECTURE

WRITINGS ON ARCHITECTURE

CIVIL AND MILITARY

c1460 TO 1640

*

a checklist of printed editions

compiled by

John Bury & Paul Breman

HES & DE GRAAF Publishers BV

HES & DE GRAAF Publishers BV

Tuurdijk 16

3997 MS ‘t Goy-Houten

The Netherlands

ISBN 90 6194 428 7

Preface

The list which follows is intended to bring up to date and to improve my *Renaissance architectural treatises and architectural books*, published in 1988 in the *Acts* of the 1981 'Conference on Renaissance architectural treatises' which had been convened by the Renaissance Institute [C.E.S.R.] of the University of Tours. The original list was more or less limited to works that could be considered to belong to the Renaissance, which in Northern Europe continued well into the 17th century, but now that this restriction no longer needs to apply we have instead adopted simple time limits: from the 1460s, or shortly before, when the first architectural books (Alberti, Filarete) began to appear in manuscript, until about 1640, which was the approximate terminal date established for the 1981 List.

In most respects we have adhered to the principles of inclusion and exclusion adopted in the earlier list, of which the following is a summary:

Included are all the post-medieval treatises on architecture known to us to have been *written* between shortly before 1460 and about 1640, regardless of when they were first *published* (works first published posthumously are marked with *), plus the one earlier work without which our period would have been very different, that of Vitruvius. We have followed previous practice by including subsequent editions of listed works up to the present day. Facsimile editions are listed in brackets directly after the edition they reproduce.

Suites of plates issued without text have been included only where they were intended directly as models or to illustrate the orders. Coverage of guide and travel books and town histories has been limited to those, mostly illustrated, which contain significant references to buildings, including, most notably, the remains of ancient Roman buildings [again because of their use as models].

Excluded also are works on geometry, mathematics, measuring, perspective and surveying, whether intended for architects or not; works on town planning and on gardens; the mere iconology of civil and military architecture; almost all technical manuals for the building trades, such as carpentry and stonemasonry; the literature of temporary structures, funeral monuments, furniture, ornament and inscriptions. The coverage of fortification literature excludes works on siegecraft. We have also omitted guides to *sacro monti*.

We owe much, of course, to other bibliographers, and in a few entries where the ground has been clearly and comprehensively covered by someone else's list we have cited that rather than merely duplicated it. This applies to the *Mirabilia Romae* and similar Roman guide books, to editions and translations of Vignola, and to the very complicated bibliography of Jan Vredeman de Vries; also, but for reason of space, to the many editions of a very slight text in Botero's *Aggiunte* to his *Ragion di Stato*.

On the other hand, while well aware of the diligent work of predecessors, we have nevertheless preferred to present our own entries on Serlio and on Vitruvius. That on Serlio supersedes the one I submitted to the Serlio Conference of 1987 at Vicenza, published in its *Acts*, Milan 1989, and reprinted almost unchanged in a recent English translation of Serlio's first five books. The Vitruvius entry had the benefit of Paul Breman's acquaintance with two very large private collections of Vitruviana.

Comparing the present list with its 1981 predecessor, the 170 author entries in the original have been increased to nearer 280. The coverage of military architecture especially is greatly extended, and the length of several of the principal entries has been notably augmented.

Despite every effort to be comprehensive, and accurate, such is the nature of bibliographical work that it can seldom be exempt from errors of omission or commission – and this list is no exception. It is hoped, however, that, excusing minor flaws, our compilation may provide a useful work of reference for bibliographers, booksellers, collectors, librarians, and all interested scholars and students of early modern architecture and architectural books.

John Bury

September 2000

ACCONCIO, Giacomo [died 1585]

Arte di munire le città Geneva 1582/3; Bassano 1796?

[Latin trsl. by the author] **Ars muniendorum oppidorum** Geneva 1583

[persistent references to this work seem to rest solely on a letter of the author to the publisher John Wolfe, 21 Dec 1562, in which he says the work is written but not yet published. No copies of either Geneva edition have ever been located, and the Bassano edition is equally doubtful]

ALBERTI, Leon Battista [1404-1472]

De re ædificatoria [ms presented to Pope Nicholas V around 1450; first ed by Angelo Poliziano] Florence 1485[=1486 - two issues]; Strasburg 1511, 1541; ed. by Geoffroy Tory, Paris 1512, 1543; critical ed., with Italian trsl., by Giovanni Orlandi and Paolo Portoghesi, Milan 1966; ed. by H. K. Lücke in **Alberti Index**, 4 vols, Munich 1975-79

[Italian version, perhaps the original, in vol. 4 of the] **Opere volgari**, ed Bonucci, Florence 1843; in **Opere volgari**, ed. C. Grayson, Bari 1960-73

[Italian trsl. by Pietro Lauro] **I dieci libri di architettura** Venice 1546

[Italian trsl. by Cosimo Bartoli; illustrated] **L'architettura** Florence 1550; Monte Regale [Mondovì] 1565; Venice 1565 [facs. Sala Bolognese 1985]; ed. by Giac. Leoni, London 1726 [only one copy known]; with copies after Leoni's plates, Bologna 1782; Rome 1784; Perugia 1804; with 30 plates by Costantino Gianni, Milan 1833; [bilingual: see English trsl. by Leoni]

[French trsl. by Jean Martin] **L'architecture et art de bien bastir** Paris 1553

[Spanish trsl. by Francisco Lozano] **Los diez libros de architectura** Madrid 1582 [facs. Oviedo 1975; Valencia 1977], 1640, 1797

[English trsl. by Giacomo Leoni from, and with, Bartoli's Italian] **The architecture** London 1726[-27] [micropublished as no 48 on reel 8 of *American Architectural Books* on Helen Parks' list, New Haven CT 1972], sheets reissued 1739; [English only] 17[53-]55 [facs. ed. by Joseph Rykwert, London 1955, 1966]

[German trsl. by Max Theuer] **Zehn Bücher über die Baukunst** Vienna 1912

[Russian trsl. by V. P. Zoubov] 1935-37

[Czech trsl.] **Deset knih o stavitelstvi** Prague 1956

[Polish trsl.] **Ksiag dziesiec o sztuca budowania** Warsaw 1960

[Italian trsl. by Giovanni Orlandi, in the critical ed. of *De re aedificatoria* by him and Paolo Portoghesi] Milan 1966

[English trsl. by Franco Borsi in] **The complete works** New York 1986

[English trsl. by Joseph Rykwert, Neil Leach and Robert Tavernor, assisted by H. K. Lücke, with intro and notes] **On the art of building** Cambridge MA 1988

[see also under PELLEGRINI]

* **Descriptio urbis Romae**, critical edition in Latin and French with commentaries in French by Martine Furno and Mario Carpo, Geneva 2000

ALBERTINI, Francesco degli [died 1517/20]

Opusculum de mirabilibus novae & veteris urbis Romae Rome 1510 [facs. in 100 copies, n.pl., c1955; another in *Five early guides to Rome and Florence*, ed. P. Murray, Farnborough 1972], 1515; Basel 1519; Bologna 1520; Lyons 1520; as first of ten texts in *De Roma prisca et nova varii auctores*, ff 1-80, Rome 1523.

Critical editions: ed. August Schmarsow, Heilbronn 1886; ed. Roberto Valentini & Giuseppe Zucchetti, *Codice topografiche della città di Roma*, vol. 1, Rome 1940

ALFARANO, Tiberio [died 1596]

* **De basilicae Vaticanae antiquissima et nova structura** [late 16th century text, with drawings and an extension project by the author, known in at least ten mss] ed. with intro by D. Michele Cerrati, Rome 1914

ALGHISI, Galasso [died 1573]

Delle fortificationi libri tre [Venice] 1570 [the often cited edition of 1575 does not exist]

[selected chapters, with notes] in Paolo Barocchi, *Scritti d'arte del Cinquecento* vol. 3, Milan-Naples 1977

ALMELA, Juan Alonso [active late 16th century]

* **Descripción de la octava maravilla del mundo** [completed 10 March 1594] ms published in *Documentos para la historia del monasterio de El Escorial* ed. Julian Zarco Cuevas, vol 6, 1962, pp 5-98

ALSTEDIUS, Joannes Henricus [c1588-1638]

Encyclopaedia Herborn 1620, 1630; Lyons 1649
[Tome VII, part 4, book 34 deals with architecture]

AMICO, Fra Bernardino [Franciscan; active 1593-1619]

Trattato delle piante & immagini de sacri edifizj di Terra Santa Rome 1609 [with plates engr. by Antonio Tempesta]; Florence 1620/19 [enlarged, and with plates engr. by Jacques Callot]

[English trsl. by T. Bellorini & E. Hoade] **Plans of the sacred edifices of the Holy Land** Jerusalem 1953

AMMANNATI, Bartolomeo [1511-1592]

* **La città. Appunti per un trattato** ed. Mazzino Fossi, Rome 1970

ANDROUET DU CERCEAU, Jacques [1510/15-1584]

Quinque et viginti exempla arcuum partim a me inventa, partim ex veterum sumpta monumentis tum Romae tum alibi etiam num extantibus Orleans 1549

Quoniam apud veteres alio structurae genere templa fuerunt aedificata Orleans 1550
[plans, elevations and sections of ancient temples, tombs and great houses, known as *Les grands temples* or, in reduced format, *Les petits temples*]

Praecipua aliquot Romanae antiquitatis ruinarum monimenta ... ad veri imitationem ... designata Paris 1550?; copies by Hieronymous Cock, same title, Antwerp 1551; [also a smaller suite as] **Operum antiquorum Romanorum ... reliquias ac ruinas** [Antwerp] 1562

Duodecim fragmenta structurae veteris commendata monumentis [after Léonard Thiry] Orleans 1550, 1565; copies by Virgil Solis, **Buechlin von den alten Gebeuen** Nuremberg, n.d.

Optices, quam perspectivam nominant viginti figuras Orleans 1551

De architectura Paris 1559

[French version] **Livre d'architecture ... contenant ... cinquante bastiments** Paris 1559, 1582, 1611

De architectura ... opus alterum Paris 1561

[French version] **Second livre d'architecture** Paris 1561

Livre d'architecture [pour les champs] Paris 1572, 1582, 1615, 1648

[facs. ed. of the three 'architecture' books , in the editions of 1559, 1561 and 1582, as] **Les trois livres d'architecture** Ridgewood NJ 1965

Le premier volume des plus excellents bastiments de France Paris 1576

Le second volume ... Paris 1579

Le premier [second] volume ... Paris 1607 [facs. Farnborough 1972], 1648; ed. H. Destailleur, plates engraved by Faure Dujarric, 1868-70; with intro and commentary by David Thomson, 1988

Petit traitte des cinq ordres de colomnes Paris 1583

Livre des édifices antiques romains [Paris] 1584

ANGELI, Paolo degli [1580-1647]

Basilicae S. Mariae Maioris ... descriptio et delineatio Rome 1621

Basilicae veteris Vaticanae descriptio Rome 1646

ANONYMOUS

* **'Bramantino' sketchbook** [Bibl. Ambrosiana] published as *Le rouine di Roma al principio del secolo XVI*, ed. Giuseppe Mongeri, Milan 1875 [200 copies], 1880 [85 copies]; see also Maria A. Phillips, *The Ambrosiana's sketchbook on the ruins of ancient Rome: its function and meaning* in Les traités d'architecture de la Renaissance, ed. J. Guillaume, Paris 1988, pp 151-167

ANONYMOUS

* **Codex Escurialensis** [compilation of drawings by more than one hand, after 1491 and before 1509; Bibl. de El Escorial] published as *Codex Escurialensis: ein Skizzenbuch aus der Werkstatt Domenico Ghirlandaios*, ed. Hermann Egger, Christian Huelsen & Adolf Michaelis, 2 vols, Vienna 1905-06 [facs. 1960]

ANONYMOUS

* **'Mantegna' codex** [late 15th or early 16th century; Kunstbibliothek Berlin MS OZ111] published as *Il codice detto del Mantegna*, ed. Luca Leoncini, Rome 1993

ANONYMOUS

* **North Italian sketchbook** [by two or more anonymous hands; Soane Museum, London] partly published in Marcel Rothlisberger, *Un libro inedito del Rinascimento Lombardo* [Palladio 7, 1957]. See also M. M. Licht, *A book of drawings by Nicoletto da Modena* [Master Drawings 8, 1970]

ANONYMOUS

* **Mellon** [formerly 'Menicantonio'] **sketchbook** [Pierpont Morgan Library, New York – formerly in the Mellon Coll.] partly published in Hans Nachod, *A recently discovered sketchbook of an intimate assistant of Bramante* [H. P. Kraus, New York, cat. Rare Books 8, 1955] and Rudolf Wittkower, *The 'Menicantonio' sketchbook in the Paul Mellon Collection* [Idea and Image, London 1978]. See Sebastian Storz in *Raffaello architetto*, exhibition catalogue, Milan 1984, pp 422-23

ANONYMOUS

* **Codice dei cinque ordini** [Bibl. Vaticana, Fondo Chigiano] published as *'I cinque ordini architettonici' e L. B. Alberti*, ed. Franco Borsi [Studi e documenti di architettura 1, 1972]

ANONYMOUS

Viazo [viaggio] **da venesia al sanctis iherusalem et al monte sinai** Bologna 1500
[contains 144 woodcuts of architectural and other subjects; the text, accompanied by different woodcuts, later appears attributed to Fra Noè, q.v.]

ANONYMOUS

Le quatrieme livre d'Amadis de Gaule Paris 1543 [three issues] - for later editions see H. Vagenay, *Amadis en français*, Florence 1906

[French version, by Nicolas de Herberay, of a romance of chivalry plausibly attributed to the 13th century Portuguese writer Joham de Lobeira and first printed in the redaction by Garcia Rodriguez de Montalvo at Saragossa in 1508. In this version a magnificent castle, known as the palace of Apolidon, on Firm Island, is described, with woodcut plan and elevation on folios A3v and A4r - the latter also appearing in Jean Martin's translation of Vitruvius. See André Chastel, *The palace of Apolidon* (Zaharoff lecture 1984) Oxford 1986]

ANONYMOUS

* **Fossombrone sketchbook** [dated between 1524 and 1538, attributed to an unidentified follower of Raphael; Bibl. Civica Passionei, Fossombrone] published as *Das Fossombroner Skizzenbuch*, ed. Arnold Nesselrath, London 1993

ANONYMOUS

* **Traité sur les ordres d'architecture** [Bibl. Nationale, Paris] published as *Un traité inédit sur les ordres d'architecture et le problème des sources du Libro IV de Serlio*, ed. Vladimir Juren [Monuments et mémoires publiés par l'Académie des Inscriptions et Belles-Lettres 64, 1981, pp 125-239]

ANONYMOUS

* Ms Bibl. Nacional Madrid, no 9.681, partly published by F. Marias & A. Bustamante as *Un tratado inedito de arquitectura de hacia 1550* in Boletin del Museo e Instituto 'Camon Aznar' 13, 1983, pp 41-57

ANONYMOUS

* **Variorum architectorum delineationes portarum et fenestrarum quae in urbe Florentiae reperiuntur** [ms of 1579-80, Bibl. Vaticana] published with intro and notes by Luigi Zangheri in *Il disegno interrotto*, Florence 1980, pp 323-71

ANONYMOUS

* **Disegni de le ruine di Roma e come anticamente erono** [formerly Dyson Perrins Collection] published as *Topographical study in Rome in 1581*, ed. Thomas Ashby, London 1916; facs. ed. by Rudolf Witkower, 2 vols, Milan 1963. Critical study of dating and authorship: Henri Zerner, *Observations on Dupérac and the 'Disegni de le ruine di Roma'* [Art Bulletin 47, 1965]

ANONYMOUS

* **Miscellanea di architettura** [ms from the 1580s attributed to Oreste Vannocci Biringucci, d. 1585, Bibl. Comunale degli Intronati, Siena] published with intro and notes by Gabriele Morolli in *Il disegno interrotto*, Florence 1980, pp 203-91

ANTIQUITATES architectonicae - see Hans BLUM

ANTONIO [Mazzone de' Liberi] **da Faenza** [c1476-1535]

***Codex Bury** [treatise of c1520 on various subjects, incl. architecture on ff 22r-66r, described and illustrated by Michael Bury, *A newly discovered architectural treatise of the early Cinquecento* in Annali di architettura 8, 1996, pp 21-42

ARCHITECTURA antiqua - see Rudolph WYSSENBACH

ARFE y Villafañe, Juan de [1535-1603]

De varia commensuración Seville 1585-87 [facs., intro by A. Bonet Correa, Madrid 1974-78; intro by F. Iñiguez, Valencia 1979], 1589?; Madrid 1675, 1736, 1763, 1773, 1795, 1806

Descripción de la custodia de la Sancta Iglesia de Sevilla Seville 1587, 1887

[English trsl. in L. Williams, *Arts & crafts of Spain*, vol. 1, pp 185-201] London 1907

ARIAS MONTANO, Fray Benito [Jeronymite; 1527-1598]

Antiquitatum Judaicarum libri IX Lyons 1593
[includes his reconstruction of the Temple of Solomon]

AVERLINO, Antonio - see FILARETE

B., le S[ieur]

Recueil de plusieurs desseins de fortifications Paris 1631, ?1639

BACCO, Enrico & **CARACCIOLO,** Cesare d'Engenio [both active early 17th century]

Nuova descrittione del regno di Napoli Naples 1629, 1644, 1646, 1671; Bologna 1977
[earler editions, as *Descrittione del regno di Napoli* by Henrico Bacco Alemanno, of which there are at least seven between 1608 and 1626, do not have the descriptions of towns and buildings added to the *Nuova descrittione*]

[Latin trsl.] **Novi descriptio regni Neapolitani** Louvain 1723

[English version trsl. and ed. by Eileen Gardiner, with intro by C. Bruzelius and R. G. Musto] **Naples, an early guide** New York 1991

BACHOT, Ambroise [active late 16th century]

Le Timon ... joinct un traicte fort utille des fortifications ... inventés par l'auteur. Paris 1587 [on p 40 dated 1579, and at end 1583]

Le gouvernail ... l'architecture de fortifications ... Melun 1598

BAEGK, Theodor [1599-1676]

Architectonica militaris defensiva Lucerne 1635

BAGLIONE, Giovanni [c1573-1644]

Le nove chiese di Roma Rome 1639; with intro by Liliana Barroero, notes by Monica Maggiorani and Cinzia Pujia, Rome 1990

BALDI, Bernardino [1553-1617]

Descrittione del Palazzo ducale d'Urbino in his *Versi e prose*, Venice 1590, pp 503-573; in Eugenio Alberi's *Tesoro della prosa italiana*, Florence 1841; and in *Versi e prose* ed. Filippo Ugolini and Filipp-Luigi Polidori, Florence 1859

[Serbo-Croat trsl by Andrea Mutujakovic, in his] *Vojvodska palaca u Urbino*, Zagreb 1992
[a reconstruction, with summaries in Italian and English]

De verborum Vitruvianorum significatione Augsburg 1612; as **Lexicon Vitruvianum** auctus & *illustratus* a Joanne de Laet, in the latter's ed. of Vitruvius, Amsterdam 1649, vol 2 pp 1-144; in Poleni's *Excercitationes*, Padua 1739-41

Scamilli impares Vitruviani explicati Augsburg 1612; in De Laet's ed. of Vitruvius, Amsterdam 1649, vol 2 pp 145-164; in Poleni's *Excercitationes*, Padua 1739-41; in Poleni & Stratico's ed. of Vitruvius, vol 1,Udine 1825, pp 239-57

BARBARO, Daniele [Monsignore; 1514-1570]

La pratica della perspettiva Venice 1568, 1569/68, 1569 [variant issues; resp. with printed title, engraved title, or woodcut title border]
[part 4, pp 129-158, deals with the orders, and with stage scenery [facs. Sala Bolognese 1980]

BARCA, Giuseppe [1595-1639]

Breve compendio di fortificatione moderna Milan 1639; Bologna 1643

BARCA, Pietro Antonio [active from c1580, died 1536 or 1639]

Avvertimenti, e regole circa l'architettura civile ... et architettura militare Milan 1620

BAROZZI, Giacinto [1535/40-c1584]

Offerta d'un nuovo modo di difendere qual si voglia fortezza Rome 1578

Seconda proposta in materia d'una nuova difesa Perugia 1581

[both are insubstantial pamphlets, and very rare]

[French trsl. by G. D. L.] **Invention admirable de défendre toutes places** Paris 1583

BAROZZI, Giacomo - see Giacomo Barozzi da VIGNOLA

BASSI, Martino [1542-1591]

Dispareri in materia d'architettura et perspettiva Brescia 1572; ed. Franc. Bern. Ferrari, Milan 1771

BELLINI, Jacopo [c1400-1470/71]

* **Les dessins de Jacopo Bellini au Louvre et au Bitish Museum** ed. Victor Golubew. 2 vols. Brussels 1908-12 [also in German as **Die Skizzenbücher**] - a projected third volume was never published

* **Jacopo Bellini, the Louvre album of drawings** ed. Bernhard Degenhart & Annegrit Schmitt. New York 1984

BELLUZZI [Bellucci. Belici], Giovan Battista [1506-1554]

Nuova inventione di fabricar fortezze Venice 1598 [two issues with different imprints but otherwise identical]
[a very corrupt text, only partly by Belluzzi, with 'particelli e fragmenti' attributed to Antonio Melloni]

* **Delle fortificazioni di terra** [ms of 1545 in the Bibl. Riccardiana, Florence] ed. with intro and notes by Daniela Lamberini in *Il disegno interrotto: trattati medicei d'architettura*, Florence 1980, pp 373-513

BENTIVOGLIO, Cornelio [1519/20-1585]

?**Discorso delle fortificazioni** Venice 1598
[this title is often referred to, but its existence has never been proved conclusively]

BERNEGGER, Mathias [1582-1640]

De fortalitiis Strasburg 1616

BERTANO, Giovanni Battista [1516-1576]

Gli oscuri et dificili passi dell'opera ionica di Vitruvio Mantua 1558

[Latin trsl. by Poleni] in his *Exercitationes*, Padua 1739-41; in Poleni & Stratico's ed. of Vitruvius, vol 1, Udine 1825, pp 277-97

BESTE, Charles de - see Charles DE BESTE

BETUSSI, Giuseppe [c1512-?1573]

Ragionamento sopra il Cathaio Padua 1573; Ferrara 1669

BINET, Etienne [S.J., 1569-1639]

Essay des merveilles de nature, et des plus nobles artifices Rouen 1621, 1622, ?, ?, 1625 [5e éd.], 1626, 1629, 1631; Paris 1632, c1639, 1644, 1646, 1657 [13e éd. - which seems to be the last]
[an encyclopedia of little interest, published under the pseudonym René François, but the chapter called *Architecture* is a thorough dictionary of architectural terms, about 40 pages, with good explanatory woodcuts]

BLUM, Hans [1520/27-c1570]

Quinque columnarum. Exacta descriptio atque deliniatio Zurich 1550 [facs. Farnborough 1967; microfilm, Fowler Coll. 52, New Haven 1979]

[German trsl.] **Von den fünff Seulen** Zurich 1550, 1554/55, 1558, 1562, 1567, 1579; as **V: Columnae: Das ist, Beschreibung unnd Gebrauch der V. Säulen,** 1596, 1627 [these two editions announce on their title-page also *hochnotwendige Architekturstucken*, which is the *Kunstrych Buch* listed below (only here is it credited to Blum, the editions of the work all appeared anonymously) and *allerley wahrhaffte*

Contrafacturen for which see under Wyssenbach - both works were in fact issued separately]; as **Nützlichs Säulenbuch** 1660, 1662, 1668, 1672; as **Grundtlicher Bericht von den fünff Sülen** Amsterdam 1612; as **Architecture nach antiquitetischer Lehr ... Die fünff Termen verordnet durch ... Rutger Kaessmann** Cologne 1644

[French trsl.] **Les cinq coulomnes de l'architecture** Antwerp 1551

[French trsl.] **Les cinq ordres des colomnes de l'architecture** Lyons 1562, 1648/49

[Flemish trsl. by Hans Liefrinck, with French trsl.] **Vande vijf colomnen van Architecture** Antwerp 1572, 1575, 1640, 1642

[Dutch trsl.] **Een constich boeck vande vijf columnen** Amsterdam 1598, ?1612, 1616/17

[English trsl. by I(ohn) T(horpe?)] **The Booke of Five Collumnes of Architecture** [with woodcut illus.] London 1601, 1608, [1620 only in Cicognara], 1635; [with engraved illus.] 1660 [facs. London 1912]; as **A description of the five orders of columnes and tearms of architecture** [with engraved illus.] [1662?], 1668, 1674, 1678 [facs. Farnborough 1967]

[Dutch and French trsl.] **Beschryvinghe van de vijf colomnen, van architecture/Description des cinq ordres de colomnes** [with additions from Dietterlin, and engraved illus.] Amsterdam 1623, 1634, 1640/41, 1647

[Russian trsl. based on Zurich 1596] Moscow 1936

Ein kunstrych Buch von allerley antiquiteten, so zum Verstand der fünff Seülen der Architectur gehörend Zurich c1560, c1580 [two issues]; as **Antiquitates architectonicae: wolbewärte Architektur-stucken, zum verstand der V. Seulen nutzlich und dienstlich** 1596; back under its original title 1627 [see notes on the last two editions under *V. Columnae*, above], 1662

several later works, under other authors' names, are simply pirated editions of Blum - so Claes Jansz **Visscher**, Beschryvinghe van de vijf Colomnen, Amsterdam 1647; Christian Martin **Anhelt**, Architectura, klaere en duydelijcke demonstratio der vijf ordens, Amsterdam 1657; Abraham **Leuthner**, Grundtliche Darstellung der fünff Seüllen, Prague 1677, and the various editions of Rütger **Kasemann** [q.v.]

BOCCHI, Francesco [1548-1613 or 1618]

Le bellezze della città di Fiorenze Florence 1591 [facs., intro by John Shearman, Farnborough 1971], 1592; enlarged by Giovanni Cinelli, 1677; Pistoia 1678

Epistola ... ruinam stragemque fractae pergamenae Florentinae testitudinis deplorantis Florence 1604

Epistola seu opusculum de restitutione sacrae testitudinis Florentinae Florence 1604

BOILLOT, Joseph [1560-1603]

Nouveaux pourtraitz et figures de termes pour user en l'architecture Langres [1592]; ed. Paulette Choué and Georges Viard, [Paris] 1995

[German trsl.] **New Termis Buch** [Strasburg] 1604

[Re-engraved version as] **Livre de termes d'animaux et leurs antipaties** Paris [Pierre van Lochom] mid-17c, [Balthasar Moncornet] late-17c, [Mariette] c1750

BORDINI, Giovanni Francesco [c1536-1609]

De rebus praeclare gestis a Sixto V. Rome 1588 [laudatory verses of no architectural interest; some of the 15 engravings show recent building projects including St Peter's]

BORROMEO, Carlo [Cardinal archbishop of Milan; 1538-1584]

Instructionum fabricae et supellectilis ecclesiasticae libri II Milan 1577; in *Acta Ecclesiae Mediolanensis*, Milan 1582 [folios 177-212], 1599; Venice 1595; Brescia 1603; Paris 1643; Lyon 1682; Bergamo 1738; Padua 1754; Milan 1843-44; ed. A. Ratti, 1892; ed. C. Castiglioni and

C. Marcora, 1952]; also in *Trattati d'arte del Cinquecento* ed. Paola Barocchi, vol 3, Bari 1962, pp 1-113, 403-6, 425-64

[Italian trsl.] Naples 1688; Milan 1823, 1952, etc.

[French trsl.] Paris 1855

[English trsl. by G. J. Wigley] London 1857

[English trsl. by Evelyn Carole Voelker] thesis Ann Arbor 1977

BOTERO, Giovanni [1544-1617]

Discorso intorno alla fortificatione [five-page essay in the *Aggiunte* to the *Ragion di stato*, Venice 1598; facs. ed. Luigi Firpo, Sala Bolognese 1990] - bibliography in Giuseppe Assandria, *Giovanni Botero*, 1928

BRANCA, Giovanni [1571-1645]

Manuale di architettura Ascoli 1629; Rome 1718, 1757; re-engraved and ed. by Leonardo de' Vegni, Rome 1772 [facs. Florence 1975], 1781, 1783, 1784, 1786; revised by Giuseppe Soli, Modena 1789

[Spanish trsl. by Manuel Hijosa] Madrid 1790

BRAY, Salomon de [1595-1664]

[introduction to De Keyser's *Architectura moderna*] Amsterdam 1631, 1641; ed. by E. R. M. Taverne, Soest 1971

BREYDENBACH, Bernhard von [c1440-1497]

[**Peregrinationes in Terram Sanctam**, otherwise listed as **Opusculum sanctarum peregrinationum ad sepulcrum Christi**, but with incipit:] **Opus transmarine peregrinationis ad venerandum et gloriosum sepulchrum dominum in Iherusalem** [illustrated by Erhard Reuwich] Mainz 1486; Speyer 1490 [with same woodcuts], 1502 [with copied woodcuts]; Wittenberg 1536

[German trsl.] [**Reise ins Heilige Land**, incipit:] **Der fart uber mer zu dem heiligen grab unsers herren** Mainz 1486 [same illustrations as Latin ed.]; Augsburg 1488 [with copies only of the small woodcuts]; Speyer, after 1502 and probably c1505

[Dutch trsl., perhaps by Reuwich; incipit:] **Heilige bevaerden over dat meer totten eerweerdighen heilighen grave ons heren** Mainz 1488 [same illustrations as the 1486 Latin]

[French adaptation by Nicolas le Huen, substituting results of his own voyage; incipit:] **La peregrination de oultre mer en terre saincte** Lyons 1488 [illustrations copied after those of 1486]; as **Le grant voiage de Jhérusalem** [with a second part added] Paris 1517, 1522

[French trsl. by Jehan de Hersin, illustrated from the original blocks] **Le saint voiage et pelerinage de la cite saincte de hierusalem** Lyons 1489/90

[abridged Czech trsl.] **Traktát o zemi svate** Pilsen 1498

[Spanish trsl. by Martin Martinez de Ampies, illustrated from the original blocks, with additional cuts] **Viaje de la tierra sancta** Saragossa 1489 [reprint, together with the translator's *Tratado de Roma*, Madrid 1974]

[German trsl. by Eliz. Geck] **Die Reise ins Heilige Land** Wiesbaden 1961

BRIANO, Giacomo [S.J.; 1589-1649]

* [A fragment of his architectural treatise survives on the versos of his drawings for buildings in northeastern Italy and in Poland; illus. catalogue by John Bury for Francesco Radaeli] **Giacomo Briano, S.J.: architecural drawings** Milan 1983 [the album now in the Getty Centre]

BULENGER, Jules César

De theatro, ludisque scenicis libri duo Troyes 1603
[chapters 6-24, ff 23v-62v, give history of Greek and Roman theatre building]

BULLANT, Jean [1510/15-1578]

Reigle generalle d'architecture des cinq manieres de colonnes Paris 1564, 1568, 1619; Rouen 1647

BUCH, Ein kunstrych, [von allerley antiquiteten] - see Hans BLUM

BUSCA, Gabriello [c1540-c1600]

Della espugnatione et difese delle fortezze Turin 1585; with *Instruttione de' bombardieri* 1598

[German trsl.] **Zwei Bücher von Bestürmung und Beschützung der Festungen** Francfort 1619

Della architettura militare Milan 1601; as **L'architettura militare** 1619

[Spanish trsl.] **Arquitettura militar** Milan 1619

CALVO, Marco Fabio [died 1527]

Antiquae urbis Romae simulachrum Rome 1527, 1532 [two issues] [facs. with postscript by Roberto Peliti, 1964]; Basel 1556, 1558; woodcuts replaced with engravings by G. B. de Cavaleri, Rome 1592

CAPO BIANCO, Alessandro [died 1610]

Corona e palma militare di artiglieria ... con una giunta della fortificatione moderna [8 pages only] Venice 1598, 1602, 1618, 1647

CAROTO, Giovanni [c1488-1563/66]

De le antiqita de Verona Verona 1560 [facs. Sala Bolognese 1976], revised 1764

Caroto's illustrations were used earlier in Sarayna [q.v.]

CASTIGLIONE, Baldassare [Count: 1478-1529]

* Report of 1519 – see RAFAELLO

CASTRIOTTO, Giacomo [Fusto. detto - 1501 or 1510-1563] - see MAGGI & Castriotto

CATANEO, Girolamo [c1500-1572]

Opera nuova di fortificare Brescia 1564; as **Libro nuovo di fortificare** 1566/7; in *Dell'arte militare libri tre* 1571; in *Dell'arte militare libri cinque* 1584, 1608

[French trsl. by Jean de Tournes] **Le capitaine** Lyons 1574, ?1589, 1593, 1600

[Latin trsl.] **De arte bellica** Lyons 1600

Nuovo ragionamento del fabbricare le fortezze Brescia 1571

[German trsl.] **Neu Gespräch, wie man Vestungen bauen solle** Eisenach 1606

CATANEO, Pietro [c1510-1569]

I quattro primi libri di architettura Venice 1554 [facs. Ridgewood NJ 1964]

L'architettura ... sono aggiunti di più il quinto, sesto, settimo, e ottavo libro Venice 1567 [facs. Sala Bolognese 1982]; critical ed. by Paola Marini, intro and notes by Elena Bassi, in *Pietro Cataneo, Giacomo Barozzi da Vignola, Trattati* ... Milan 1985, pp 163-498

CAVALCA, Alessandro [died 1645]

Essamine militare ... con l'aggiunta ... di fortificationi Venice 1620

[the *Aggiunta* is not in the Essamine's first edition of 1616]

CELLINI, Benvenuto [1500-1571]

* **Discorso dell'architettura** [ms of uncertain date] in Jacopo Morelli, *I codici manoscritti volgari della Libreria Naniana*, Venice 1776; in vol 3 of Cellini's **Opere** Milan 1811; in *Vita di Cellini* ed. Gio. Palamede Carpani, vol 3, Milan 1821, pp 190-98; in *Vita di Cellini* ed. Francesco Tassi, vol 3, Florence 1829; in *I trattati di Cellini* ed. Carlo Milanese, Florence 1857, pp 220-28; in *Vita di Cellini* ed. A. J. Rusconi and E. Valeri, Rome 1901, pp 795-99; in *I trattati di Cellini* ed. L. De Mauri, Milan 1927, pp 255-67; etc.

[French trsl. in *Oeuvres de Cellini*, vol 2] Paris 1847

CIAPPI, Marc'Antonio [active late 16th century]

Compendio delle heroiche, et gloriose attioni, et santa vita di Papa Greg. XIII [Con le Figure tratte dal naturale delli Collegij, Seminarij, & altre Fabriche fatte da lui] Rome 1596

COCK, Hieronymus [Jeroen, c1510-1570] – see Jacques ANDROUET du Cerceau, and Vincenzo SCAMOZZI

COECKE van Aelst, Pieter [1502-1550]

Die inventie der colommen met haren coronementen ende maten Antwerp 1539 [only two copies known; facs. ed. Rudi Rolf, Amsterdam 1978]

COLONNA, Fra Francesco [Dominican; reputedly born 1432/33; died 1527 not 1476/77]

Poliphili hypnerotomachia Venice 1499 [facs. London 1904; with separate intro by G. D. Painter, London 1963; Milan 1963; reduced Ridgewood NJ 1969; New York 1976; with intro by Peter Dronke, Saragossa 1981]; as **La hypnerotomachia di Poliphilo, cioè pugna d'amore in sogno** 1545; as vol 2, ed. by Giovanni Pozzi, of Maria Teresa Casella, *Francesco Colonna, biografia e opere*, Padua 1959; critical ed. by G. Pozzi and L. A. Ciapponi, Padua 1964, 1980; ed. Marco Ariani and Mino Gabriele, with extensive bibliography, Milan 1998

[French trsl. by an 'eques Meltensis', perhaps Robert De Lenoncourt] **Hypnerotomachie, ou Discours du songe de Poliphile** Paris 1546, 1553/54, 1561; abbreviated by B. Guégan, 1926; critical ed. by Gilles Polizzi, Paris 1994

[English trsl. of the first 16½ chapters of book I by 'R. D.'] **Hypnerotomachia. The strife of love in a dreame** London 1592 [facs. New York & Amsterdam 1969; with intro by Lucy Gent, Delmar NY 1973; New York 1976]; ed. by Andrew Lang, London 1890

[French trsl., ed. F. Béroalde de Verville] **Le tableau des riches inventions** Paris 1600 [three issues thus, the third certainly after 1610], 1657

[French abbrev. trsl. by J. G. Legrand, with intro; 2 vols] Paris 1804; Parma 1811

[French trsl. by Claude Popelin, with intro and notes; 2 vols] **Le songe de Poliphile ou Hypnerotomachie** Paris 1883 [facs. Geneva 1971, 1994]

[Spanish trsl. with intro and notes by Pilar Pedraza] Murcia 1981

Fac-similes of one hundred and sixty-eight woodcuts in Hypnerotomachia Poliphili, Venice 1499, with intro and descriptions by J. W. Appell, London 1888, 1893

[French trsl. by Albert-Marie Schmidt] Paris 1963

[English trsl. by Joscelyn Godwin] London 1999

CONER, Andreas [died 1527] – see Bernardo della VOLPAIA

CONTRAFACTUREN, Wahrhaffte - see Rudolph WYSSENBACH

CORNARO, Alvise [1475 or 1484-1566]

* **Due trattati di architettura e saggio sul duomo di Padova** [two mss, Bibl. Ambrosiana, Milan, cod. A 71 inf. and R 124 sup., representing a first and second redaction, c1550 and c1555] ed. G. Fiocco in Atti dell' Accademia Naz. dei Lincei, Memorie, serie 8, vol 4, fasc. 3, Rome 1952, pages 207-222; in G. Fiocco, *Alvise Cornaro*, Vicenza 1965, pp 156-167; excerpts and bibliography in Paola Barocchi, *Scritti d'arte del Cinquecento* vol 3, Milan-Naples 1977, pp 3134-61 and 3561-63; ed. P. Carpeggiani as *Scritti sull'architettura*, Padua 1980; ed. C. Semenzato in *Pietro Cataneo, Giacomo Barozzi da Vignola, Trattati*, Milan 1985, pp 77-113

COSTAGUTI, Giovanni Battista [died 1625]

Architettura della basilica di S. Pietro in Vaticano Rome 1620, 1684

[a 1620 edition is mentioned on the title-page of the 1684, but no copy has ever been located. The British Architectural Library lists its copy as being of 1620, but this is a made-up set of plates only which may well be of the 1684 impressions; the set also includes several plates dated 1631 and 1640 or bearing the mid-17th century imprint of Giov. Battista de Rossi]

the plates, by Martino Ferrabosco, were issued separately, 1684, 1812

CRONACA - see Simone del POLLAIUOLO

DAMANT, Le Sieur

Manière universelle de fortifier Brussels 1630

DANTI, Vincenzo [1530-1576]

Il primo libro del trattato delle perfette proporzioni Florence 1567 [reprinted in Paolo Barocchi, *Trattati d'arte del Cinquecento* vol 1, Bari 1960, pages 207-69]

DE BESTE, Charles [active late 16th century]

* **Architectura. Dat is constelicke bouwijnghen uijt die Antijcken ende modernen** [ms II 7617, Bibl. Royale, Brussels, written c1595-99] partly quoted and reproduced by C. van den Heuvel in Handelingen van het Genootschap voor Geschiedenis 131, 1994, pp 65-93, and Bulletin van de Kon. Nederl. Oudheidkundige Bond 94, 1995, pp 11-23

DE L'ORME, Philibert [1514-1570]

Nouvelles inventions pour bien bastir et a petits fraiz Paris 1561, 1568, 1576 [sheets reissued 1578]; and as books 10 and 11 of the **Architecture** [see below] 1626, Rouen 1648

Le premier tome de l'architecture Paris [variant issues dated] 1567 or 1568; 1576; with *Nouvelles inventions* as **Architecture, oeuure entiere** [with 28 additional woodcuts] 1626 [facs. with intro by C. Nizet, Paris 1894], Rouen 1648 [facs. Ridgewood NJ 1964; with intro by Geert Bekaert, Brussels 1981; with intro by J. M. Pérouse de Montclos, Paris 1988]
[the promised second volume never appeared]

DEMONTIOSUS, Ludovicus [Louis de Montjosieu; active second half of 16th century]

Gallus Romae hospes. Ubi multa antiquorum monimenta explicantur. Rome 1585

DESCRIZIONE di Roma antica [moderna] - see Pompilio TOTTI

DIETERICH, Conrad [1575-1639]

Discursus politicus de munitionibus et propugnaculis Giessen 1608 [thesis], 1620; Frankfurt 1643; in J. Chr. Lippold, *Arbor consanguinitatis*, [Merseburg] c1650, pp 37-110

[German trsl. by Johann Philip Ebel] **Politischer Discurs von Festungen** Giessen 1620; as **Politischer Tractat von Festungen** Marburg 1641

DIETTERLIN, Wendel [Wendelin Grapp - 1550/51-1599]

Architectura und Ausstheilung der V. Seulen. Das erst Buch Stuttgart 1593 [but issued Strasburg 1594]

[same in Latin and French] **Architectura de quinque columnarum ... Liber I** Strasburg 1595

Architectur von Portalen ... Das annder Buch Stuttgart 1593 [but issued Strasburg 1594]

[same in Latin and French] **Architectura de postium ... Liber II** Strasburg 1594/95

[both works, enlarged and arranged in five books] **Architectura** Nuremberg 1598[=1599], separate issues in Latin-and-French and German [facs. Liège (French) 1861, (German) 1862; with intro by Hans Gerhard Evers, Darmstadt 1965; with intro by Adolf K. Placzek, New York 1968; with intro by Erik Forssmann, Brunswick 1983], 1655

DILICH, Wilhelm [Schäffer, called - 1571/72-1650]

Kriegsbuch Cassel 1608; ed. Joh. Wilhelm Dilich, Frankfurt 1689, 1718

Peribologia oder Bericht von Vestungs gebewen Frankfurt 1640 [facs. Unterscheidheim 1971]

[Latin trsl.] **Peribologia seu muniendorum locorum ratio** Frankfurt 1641

* **Wilhelm Dilichs Federzeichnungen** kursächsischer und meisanischer Ortschaften aus den Jahren 1626-1629, ed. Paul E. Richter & Christian Krollmann, Dresden 1907

DILICH, Johann Wilhelm [Schäffer, called - 1600-c1660]

Kurtzer Unterricht wie ... mann einen ... Platz fortificieren kann Frankfurt 1642

DONATO, Alessandro [1584-1640]

Roma vetus ac recens utriusque aedificiis ... expositis Rome 1638/39, 1648; somewhat enlarged 1662, 1665; with new engravings by Jan van Munnickhuysen, Amsterdam 1694/95; then again Rome 1725, 1738

DOSI[O], Giovanni Antonio [1533-after 1610]

Urbis Romae aedificiorum reliquiae Florence 1569 [facs. with intro by Franco Borsi, Rome 1970]; as **Varie antichità di Roma** Rome 1640

* **Il libro delle antichità,** in *Giovanni Antonio Dosio: Roma antica e i disegni di architettura agli Uffizi*, ed. Franco Borsi, Cristina Acidini, et al., Rome 1976, pages 27-131

* Writing to Niccolò Gaddi in 1574 Dosio expressed his intention to compose a treatise on architecture [Bottari-Ticozzi, *Lettere* vol 3, Milan 1822, p 301]. Elements of this intended treatise have been assembled in *Giovanni Antonio Dosio* [see above] pages 167-207

DU CERCEAU, Jacques Androuet - see ANDROUET DU CERCEAU, Jacques

DÜRER, Albrecht [1471-1528]

Etliche underricht, zu Befestigung der Stett Schloss und flecken Nuremberg, October 1527 [two issues - minor corrections only; facs. Unterscheidheim 1969; in critical ed. by Alvin E. Jaeggli, Zurich 1971; facs. with intro and notes by Martin Biddle, Farnborough 1972], December 1527, ?1530, ?1538; Arnhem 1603/4; Berlin ?1803, 1823; with Italian trsl., intro and notes by G. M. Fara, Florence 1999

[Latin trsl. by Joh. Camerarius, i.e. Joachim Kammermeister] **De urbibus, arcibus, castellisque condendis** Paris 1535

[French trsl. by Evreux, intro by A. F. Ratheau] **Instructions sur la fortification des villes bourgs et châteaux** Paris 1870

DU PERAC, Etienne [1525 or 1535-1604]

I vestigi dell'antichità di Roma Rome 1575 [facs. as **The antiquities of Rome engraved in perspective and described**, Romford 1977], 1600, 1621?, 1639, 1653, 1671, 1680, 1773; as **Le cospicue e meravigliose fabriche degli antichi Romani** 1709; Utrecht 1621; reduced re-engravings by Marcus Sadeler, again as **Vestigi** Prague 1606, itself copied Rome [1660]

DU PRAISSAC

Les discours militaires Paris 1614 [two issues: one has 1612 at end, the other is called 'dernière éd.'], 1617, 1618, ['la seconde éd.'] 1622, 1623; Rouen 1625, 1628, 1636; Paris 1638, 1738
[the very short chapter four treats of fortification]

[Dutch trsl. by Zacharias Heyms] **Crychs-handelinge** Amsterdam 1623, 1635

[German trsl.] **Handbüchlein** Leipzig 1637

[English trsl. by I(ohn) C(ruso)] **The art of warre** Cambridge 1639, re-issued with new title-page 1642
[fortification chapter here on pp 31-47]

DURET, Noël [c1590-c1650]

Traité de la géométrie et de fortifications Paris 1643

EBELMANN, Johann Jakob [active early 17th century]

Seilenbuch Cologne 1611

see also under Jakob GUCKEISEN

ERRARD de Bar-le-Duc, Jean [1554-1610]

La fortification reduicte en art Paris 1600, 1604; ed. Alexis Errard 1619/20, 1619/22; Frankfurt 1604, 1617

[German trsl.] **Fortificatio** Frankfurt 1604; Oppenheim 1620; as **Kurtze Anweisung** Mömpelgard [Montbéliard] 1675

ESTEVAN, Martin [active early 17th century]

Compendio del rico aparato y hermosa architectura del templo de Salomon Alcalà de Henares 1615

EYLEND von Bellisieren, Martin

Modus artis fortificatoriae Belgicus, Niederländisch Festung Bawen Dresden 1624, 1630, 1632

FABRE, Jean

Les pratiques sur l'ordre, et règle de fortifier Paris 1629

FELINI, Pietro Martire [1565-1613]

Trattato nuovo delle cose maravigliose dell'alma città di Roma [without illustrations] Rome 1610, 1650; [illustrated] 1610 [facs. with note by S. Waetzoldt, Berlin 1969], 1615, 1625

[French version] **Les merveilles et antiquitez de la ville de Rome** Liège 1631; Douai 1639

[Spanish trsl.] **Tratado nuevo** Rome 1610, 1619, 1651

FERRABOSCO, Martino [died 1629] - see Giovanni Battista COSTAGUTI

FIAMMELLI, Giovanni Francesco [1565-1613]

Il principe difeso nel quale si tratto di fortificazione ... Rome 1604

FILARETE, Antonio di Pietro Averlino, detto il [1400/10-1469/79: the earlier dates more probable]

* **Trattato di architettura** [probably completed 1464; extant in two ms copies: Magliabecchiana and Bibl. Naz., Florence] partial ed. in German by Wolfgang von Oettingen, Vienna 1896; Piero de' Medici's copy, the Florence ms, which has an additional 25th book on Medici buildings, ed. with English trsl. by John R. Spencer, *Filarete's treatise on architecture* New Haven 1965; critical ed., with variant readings, by Anna Maria Finoli and Liliana Grassi, 2 vols, Milan 1972

[Spanish trsl. by Pilar Pedraza, with intro] Vitoria 1990

FLAMAND, Claude [c1570-1626]

Le guide des fortifications Montbéliard 1597, [1611]

[German trsl. by Hans Wieland] **Gründtlicher Underricht, von Auffrichtung und Erbawung der Vestungen**, [not Basel but] Mömpelgard [Montbéliard] 1612

FLORIANI, Pietro Paolo [1585-1638]

Diffesa et offesa delle piazze Macerata 1630; Venice 1654

FLUDD, Robert [1574-1637]

De arte militari in his *Utriusque cosmi ... historia*, vol 2, pt 2, Oppenheim 1619; Frankfurt 1624
[the work includes a section on mnemonic architecture]

FONTANA, Domenico [1543-1607]

Della trasportatione dell'obelisco vaticano Rome 1590; augmented with a **Libro secondo** recording executed buildings, Naples 1604 [facs., intro by Adriano Carugo, Milan 1978]
[the plates of the transportation and erection of the obelisk had a long after-life in Carlo Fontana's and Nic. Zabaglia's works on the Vatican]

[Spanish trsl. by Jaime Melgar with parallel English trsl. by Margaret Clark, intro by José Calavera] Madrid 1974

FORM UND WEIS zu bauwen - see Hans van SCHILLE

FRANCART [or FRANCQUART], Jacques [1582/83-1651]

Premier livre d'architecture Brussels 1616/17

FRANCESCO DI GIORGIO MARTINI [1439-1501]

* **Trattato di architettura civile e militare** [written after 1482] ed. Cesare Saluzzo, intro Carlo Promis, 2 vols and atlas, Turin 1841; with illus. from mss, in Corrado Maltese & Livia Maltese Degrassi, **Trattati di architettura, ingegneria e arte militare**, Milan 1967; ed. P. C. Marani, Florence 1980; ed. M. Mussini, *Il Codice estense restituito*, Parma 1991

[all available further information can be found in Gustina Scaglia, *Francesco di Giorgio, checklist of mss and drawings, autograph and copies 1470-1687*, Bethlehem PA 1992]

FRANCINI, Alessandro [c1570-1648]

Livre d'architecture Paris 1631 [facs. Farnborough 1966=1967], 1640

[English trsl. by Robert Pricke] **A new book of architecture** London 1669

FRANCISCUS à Dort

Beschreibung und Abcontrafeyung des ... Gebewdes ... genandt das Closter S. Laurentii gelegen in Escuriali Hamburg 1597 [reproduced in J. B. Bury, *Early printed references to the Escorial*, in *Iberia: studies in honour of H. V. Livermore*, ed. R. O. W. Goertz, Calgary 1985, plates I-VI]

[broadsheet, 43.5x63 cm, with woodcut low oblique aerial view of the monastery from the West, after Herrera, q.v.]

FRANÇOIS, René - see Etienne BINET

FRANZINI, Girolamo [1537-1596]

Le cose miravigliose dell'alma città di Roma Venice 1588; Rome 1595, 1675

Antiquitates Romanae urbis Rome 1589, 1599

[this and the following two titles are collections of the woodcut illustrations used in his (and many later) Rome guides, augmented with many which otherwise remain unpublished]

Templa Deo et Sanctis eius Romae dicata Rome 1589, 1599

Palatia Romanae urbis Rome 1596, 1599

FREITAG, Adam [1602-1664]

Architectura militaris nova et aucta oder Newvermehrte Fortification Leyden 1631, 1635; Amsterdam 1641/42, 1665

[French trsl.] **L'architecture militaire** Leyden 1635; Paris 1639/40, 1660, 1668

FRÖNSPERGER, Leonhardt [1520?-1575]

Bauw-Ordnung. Von Burger und Nachbarlichen Gebeuwen Frankfurt 1564, 1567

Kriegsbuch ... Dritter Teil: Von Schantzen und Befestigungen. Frankfurt 1573, 1596
[parts 1 and 2 have different subjects]

Vonn Geschütz und Fewerwerck ... Das ander Buch. Von erbawung, erhaltung ... der wehrlichen Bevestungen Frankfurt 1557, 1564

FULVIO, Andrea [c1470-1527]

Antiquaria urbis Rome 1513; much augmented by Girolamo Ferrucci, as **Antiquitates urbis** [1527]; as **De urbis antiquitatibus libri quinque** 1545

[Italian trsl. by Paulo dal Rosso] **Opera della antichità della città di Roma** Venice 1543; augmented, as **L'antichità di Roma** 1588

FURTTENBACH, Josef [1591-1667]

Newes itinerarium Italiae Ulm 1626 [facs. Hildesheim 1971]
[notes on his early years in Italy, when his exposure to Italian architecture caused him to change his profession from merchant to architect]

Architectura civilis Ulm 1628 [facs. Hildesheim 1971]

Architectura martialis Ulm 1630 [facs. Hildesheim 1975]

Architectura universalis Ulm 1635 [facs. Hildesheim 1975]

Architectura recreationis Augsburg 1640 [facs. Hildesheim 1971]

Architectura privata Augsburg 1641 [facs. Hildesheim 1971]

GALILEI, Galileo [1564-1642]

* **Breve instruzione all'architettura militare** [and another, slightly different, version titled **Trattato di fortificazione**, both dating from 1592-93, ms copies in public libraries of Florence, Milan, Rome and Turin] published in vol 20 of *G. Galilei: le opere*, ed. A. Favaro, Florence 1890-1909, and in vol 2 pp 7-146 of the new edition, ed. Garbasso & Alberti, Florence 1929-39

for Galileo's utilization of text and illustrations of Bernardo Puccini's *Trattato* [q.v.] see Daniela Lamberini, *Il principe difeso*, Florence 1990, pp 136-38

GALLACINI, Teofilo [1564-1641]

* **Trattato sopra gli errori degli architetti** [written 1621] Venice 1767 [facs., together with Antonio Visentini's *Osservazioni che servono di continuazione al trattato di Teofilo Gallacini* (Venice 1771), Farnborough 1970, the two works also issued separately; Sala Bolognese 1989]

a detailed summary of the Gallacini treatise is given in G. della Valle, *Lettere sanesi* vol 3, 1786

GAMUCCI, Bernardo [active second half of 16th century]

Libri quattro dell'antichità della città di Roma Venice 1565 [quarto, all others octavo]; as **Le antichità della città di Roma** Venice 1569, 1580, 1588

GEMÄLT, Wunderbarliche kostbare – see Rudolph WYSSENBACH

GENTILHâTRE, Jacques [1578- after 1623]

* Album of architectural drawings published in *Catalogue of the drawings collection of the R.I.B.A.: Jacques Gentilhâtre*, ed. Rosalys Coope, Farnborough 1972

GENTILINI, Eugenio [1529-c1603]

Instruttione de' bombardieri ... Et un discorso intorno alle fortezze Venice 1592; as **Istruttione di artiglieri** 1598, 1606; as **Il perfeto bombardiero** 1626; as **Pratica di artiglieria** 1641

GIL DE HONTAÑON, Rodrigo [c1500-1577]

* **Tratado de la composizion de los te[m]plos**, in Simon García, *Compendio de architectura y simetria de los te[m]plos co[n]forme a la medida del cuerpo humano con algunas demo[n]straziones de geometria* [ms dated 1681]. This *Tratado*, comprising the first six [or at least the first four] of García's *Compendio*, has been published by E. Mariátegui in Arte de España 7, Madrid 1868; by José Camón Aznar in *Arquitectura por Simón García*, Salamanca 1941; and by Manuel

Pereda de la Reguera in *Antologia de escritores y artistas montañeses* 20, Santander 1951. A complete facsimile, with transcript, and lexicon of García's *Compendio*, has been published by the Escuela Nacional 'Manuel del Castillo Negrete', Churubusco [Mexico] 1979, with introductory articles by Carlos Chanfón Olmos and A. Bonet Correa

GILLES, Pierre [1490-1555]

De topographia Constantinopoleos et de illius antiquitatibus Lyons 1561/62 [facs. Athens 1967]; in *Topographia lib. IV*, Leyden 1632

[English trsl. with index by John Ball] **The antiquities of Constantinople** London 1729; with intro by Ronald G. Musto, New York 1988

GIORGI, Fra Francesco [Franciscan; c1453/60-1540]

* **Promemoria per San Francesco della Vigna** [ms, now lost, first printed in] Gianantonio Moschini, *Guida per la città di Venezia*, Venice 1815, I:i pp 55-61; ed. with intro and notes by Licisco Magagnato in *Pietro Cataneo, Giacomo Barozzi da Vignola, Trattati*, Milan 1985, pp 1-18, 136-38

[English trsl.] in Rudolf Witkower, *Architectural principles in the age of humanism* London 1949, appendix I

GIOVANNOLI, Alò [active late 16th and early 17th century]

Roma antica Rome 1616, [with 143 plates] 1619; partly reprinted [or existing stock re-used] as **Vedute degli antichi vestigi di Roma … comprese in rami 106**, Rome c1750 [no two copies identical]; 19 of the plates reproduced in Alfonso Bartoli, *Cento vedute di Roma antica*, Florence 1911

GONZALEZ DE MEDINA BARBA, Diego [active 1565-1600]

Examen de fortificación Madrid 1599, ?1609

GRAPALDI, Francesco Maria [1465-1515]

De partibus aedium Parma c1494, 1501, 1506; Strasburg 1508; Paris 1511; with Life by Andrea Albio, Parma 1516; Venice 1517; Turin 1517; Basel 1533, 1541; Lyons 1535; Dordrecht 1638

[French trsl. by Jean Maure] Montauban c1520
[we have not been able to locate a copy, and are inclined to doubt the book's existence]

GRECO, El [Domenikos Theotocopoulos, known as - 1541-1614]

* Writings on architecture by El Greco are referred to by Francisco Pacheco who visited the Cretan painter at Toledo in 1611 [*Arte de la pintura* 3:9]. El Greco's son, Jorge Manuel, recorded in 1621 that he had helped his father with a notable [*insigne*] work on all aspects of architecture, with which the latter had been occupied continuously over many years – probably the 'zinco libros di arquitectura manusescriptos, el uno con trazas' in the inventory of Jorge Manuel's effects taken in that same year. This ms work has not survived, but we do have El Greco's extensive annotations in his copies of

1) the Venice 1556 edition of Barbaro's Vitruvius [Bibl. Nacional, Madrid] transcribed with commentary in F. Marías & A. Bustamante, *Las ideas artísticos de El Greco*, Madrid 1981
2) the Florence 1568 edition of Vasari's Lives [De Salas Coll., Madrid] transcribed with commentary by X. de Salas & F. Marías in *El Greco y el arte de su tiempo*, Madrid 1992
3) the Venice 1566 quarto collected edition of Serlio [Bibl. Nacional, Madrid] noticed in F. Marías, *El Greco, pintor extravagante*, Madrid 1997, pp 188 and 303 note 13

GRINGALET, Giano

Disputatio de fortalitiis Strasburg 1617

GROOTE, Alexander von [died 1621]

Neovallia. Dialogo Munich [not Monaco, as in some bibliographies] 1617

[German trsl.] **Neue Manier mit wenigen Kosten Festungen zu bauen** Munich 1618

GROSS, Huldreich

Kriegsbaw Leipzig 1632

GUCKEISEN, Jakob [active early 17th century] & Johann **EBELMANN** [q.v.]

Seilen Buch darinnen derselben Grunt, Theilung, Zieradt undt gantze Volkommenheit vorgebildet wird Cologne [1598]

HASEMANN, Henning [active first quarter 17th century]

Synopsis architecturae Oder: Summarischer Begriff der Baw-Kunst
Frankfurt 1626

HEEMSKERCK, Maerten van [1498-1574]

* **Die römischen Skizzenbücher des Marten van Heemskerck** [ms in the Kupferstichkabinett, Berlin] ed. Christian Huelsen and Hermann Egger, 2 vols, Berlin 1913-16

HEMMINGER, Sebastian

Kurtzer summarischer Bericht was … bey legung der ersten Stein zu dem vorhabenden Gebäw einer neuen Kirchen … fürgangen
Regensburg 1627

HERRERA, Juan de [c1530-1597]

Sumario y breve declaracion de los diseños y estampas de la fabrica de San Lorenzo el Real del Escurial [with 12 engravings, on 11 plates, of Herrera's own drawings engraved by Pierre Perret, 1583-89] Madrid 1589 [facs. accomp. L. Cervera Vera, *Las estampas y El Sumario de El Escorial por Juan de Herrera*, Madrid 1954 - here a second issue of the *Diseños* is recorded, with the date of 1619; another facs. Valencia 1978; and in the exhibition catalogue *El Escorial en la Biblioteca Nacional*, section C: *El Escorial, historia de una imagen: estampas y dibujos*, by Elena Santiago Paez and J. M. Magariños, Madrid 1985, pp 223-316]

The twelve engravings reproduced, with the Sumário captions in German trsl., by C. v. d. Osten Sacken, *San Lorenzo el Real de el Escorial: Studien zur Baugeschichte und Ikonologie*, Mittenwald-Munich 1979. Nine of the engravings, with their captions, are reproduced in George Kubler, *Building the Escorial*, Princeton 1982

A copy of Diseño VII [bird's-eye view from West] was published in Franciscus à Dort, *Beschreibung und Abcontrafeyung* [q.v.]

Other copies of the Diseños, from 1590 [in Perez de Mesa's *Grandezas de España*] to 1773 [in Antonio Ponz's *Viage de España* 2] are recorded in detail in E. Santiago Paez & J. M. Magariños, op. cit., pp 253-311, with very few exceptions [one being Alain Manesson Mallet, *La description de l'univers*, Paris c1684, European section ch. 12, p. 161]

Institucion de la academia real mathematica Madrid 1584 [facs. of only known copy – Bibl. Mazarine, Paris - with intro by José Simon Diaz and Luis Cervera Vera, Madrid 1995]
[on ff 15r-17r Herrera specifies the qualifications required for the training of architects and fortification engineers]

HERTFELDER, Bernhard [1587-1664]

Basilica SS. Udalrici et Afrae Augsburg 1627

[German trsl. by Roman Kistler] **Basilica das ist Herrliche Kirchen ...** Augsburg 1712

HOLANDA [or Olanda], Francisco de [1516/17-1584]

* **Os desenhos das antigualhas que vio Francisco d'Ollanda**, ed. Elias Tormo, Madrid 1940

* **Da pintura antigua** [ms of 1548, surviving only in a copy of c1790] published by Joaquim de Vasconcellos in the periodical A Vida Moderna 12-14, Oporto 1890-92, and subsequently in book form with introduction and notes: *Da pintura antigua: tratado de Francisco de Hollanda* Oporto 1918, 1930. Another edition was edited by Angel González García, Lisbon 1983.
[Livro 1, cap. 43 of Holanda's treatise is devoted to architectural theory, or 'pintura architecta']

*.[ms Spanish trsl. of 1563, in library of Academia de Belas Artes, Madrid, by the painter Manuel Denis from Holanda's original ms] published with intro and notes by P. J. Sánchez Cantón, **De la pintura antigua por Francisco de Holanda** Madrid 1921

* **Da fabrica que falece ha cidade de Lysboa** [ms of 1571] facs. in Jorge Segorado, *Francisco d'Ollanda*, Lisbon 1970, pp 67-130. Previously published, with intro and notes but without illustrations, by Joaquim de Vasconcellos in Archeologia Artistica 6, Oporto 1879; and, with its illustrations, by Vergilio Correia in Archivo Español de Arte y Arqueologia, Madrid 1929

HONDIUS, Hendrik [the elder - 1573-1650]

Les cinq rangs de l'architecture ... Avec encore quelques belles ordonnances d'Architecture ... Inventees par Iean Vredeman Frison, Amsterdam 1617, 1620; as **L'architecture** 1628, 1651, 1662 [despite the title, even the main work, on the orders, is by Vredeman]

[German trsl.] **Architectur oder Bawmeisterschafft ...** Mit beygefügter schöner Underweisung ... durch Johann Friedmann, Amsterdam 1628

[Dutch trsl. ed. by Samuel Marolois] **Architectura** Amsterdam 1638

Korte beschrijvinge ... der fortificatie The Hague 1624

[French trsl. by A. G. S.] **Description et brève déclaration** The Hague 1625

see also the note under Johan VREDEMAN de Vries

HUME, James [active first half 17th century]

Les fortifications françoises Paris 1634

IVE, Paul

The practise of fortification: wherein is shewed the manner of fortifying in all sorts of scituations London 1589 [facs. Amsterdam and New York 1968; with intro by Martin Biddle, London 1972], 1597

KASEMANN, Rütger [active 1615-1644]

Architectura. Lehr Seiulen Buch Cologne 1615; as **Seilen Bochg** 1616; as **Architectur, nach Antiquitetischer Lehr** 1630, 1644, 1653 [but preface dated 1659], 1659, 1664
[this work is really just another version of Blum's, q.v.]

[French version] **Livre d'architecture** Paris 1622

KEYSER, Hendrik de [1565-1621]

Architectura moderna Amsterdam 1631, 1641; ed. with intro by E. R. M Taverne, Soest 1971

KRAMMER, Gabriel [died c1608/09]

Architectura Von den funf Seulen Prague 1599/1600 [facs. Frankfurt 1895], 1606, 1608/09; reversed copies Cologne 1610/11

KRÖLL von Bemberg, Georg Günther

Tractatus geometricus et fortificationis Arnhem 1618
[3 vols, of which only the last deals with fortification]

LABACCO, Antonio [c1495- after 1587]

Libro appartenente a l'architettura nel qual si figurano alcune notabili antiquita di Roma Rome 1552 [at least three issues: one by the author, two by Blado with or without a dedication], 1557 [two issues by the author], 1559 [four issues, also by the author; facs. of one of these, with notes by Arnaldo Bruschi, Milan 1992]; with copied, often reversed plates, Venice 1567, 1570, 1576, 1581, 1584; partly with the original engravings, Rome [several between 1568 and 1574 by Lafrery, including one dated] 1572; 1640, mid-17c, 1672 [these three by De Rossi]; 1773

LANTERI, Giacomo [died 1560]

Due dialoghi … del modo di disegnare le piante delle fortezze secondo Euclide; et del modo di comporre i modelli Venice 1557, ?1559; Rome 1583

Due libri del modo di fare le fortificationi di terra Venice 1559; with Lupicini & Zanchi, 1601

[Latin trsl.] **De modo substruendi terrena munimenta** Venice 1563; as **De subtilitate** [under pseudonym-anagram Iolante Rabico] 1571

LAPARELLI, Francesco [1521-1570]

* **Visita e progetti di miglior difesa in varie fortezze ed altri luoghi dello Stato Pontifico** [ms of 1562] transcr. and ed. by Paolo Marconi, Cortona 1970

LA TREILLE, François de [died 1578] - see Giovanni Battista ZANCHI

LAURO, Giacomo [1584-1637]

Antiquae urbis splendor [Antiquitatum urbis liber secundus; Antiquae urbis splendoris complementum: parts I-III] Rome 1612-13-15 [but part III also issued with 10 additional plates dated 1616. Facs. with intro by Cecilia Cansetti, notes to the plates by Cristina Falcucci, as *Meraviglia della Roma antica*, Rome 1992], 1621; with notes also in Italian, German and French, 1625; with part IV, **Antiquae urbis vestigia quae nunc extant** added, 1628, 1630; ed. by 'Giovanni Alto' [= Hanns Rudi Gross] again as **Antiquae urbis splendor** Rome 1637, also as **Splendor dell'antica e moderna Roma** 1641 and perhaps later; as **Romanae magnitudinis monumenta** 1699

LECHUGA, Cristóbal [born 1557]

Discurso ... de la artilleria ... Con un tratado de fortificación Milan 1611 [the Trattado occupies pp 239-273]; as **Tratado de la artilleria y de fortificación** Madrid 1990

LELLO, Gio. Luigi [active last quarter of 16th century]

Descrizione del real tempio e monastero ... di Monreale Palermo 1588, 1596; with notes and additions by Michele del Giudice, 1702

LE MUET, Pierre [1591-1669]

Maniere de bastir, pour touttes sortes de personnes Paris 1623 [facs. Aix-en-Provence 1981]; enlarged, as **Manière de bien bastir**, 1647, 1663-4 [facs., intro Anthony Blunt, Farnborough 1966, 1972; with intro and notes by Claude Mignot, Paris 1981], 1681; fully engraved version, c1723

[as part 2 of Le Muet's Palladio trsl.] Paris 1645, 1647, 1764; Amsterdam 1682

[English trsl. by Robert Pricke] **The art of fair building** London 1670; with a second volume added 1675, 1679

LEONARDI, Giangiacopo [Count of Montelabbate; 1498-1562]

[Index only to an unpublished *Libro delle fortificationi de' nostri tempi*, on pp 39-40 of Barbaro's edition of Vitruvius] Venice 1556
[this passage is not repeated in later editions of Barbaro's Vitruvius]

* **Libro delle fortificationi de' nostri tempi** [ms dated Nov. 1553, Bibl. Oliveriano, Pesaro] ed. with intro by Tommaso Scalesse in Quaderni dell'Ist. di Storia dell'Archit., facs 115-126, Rome 1975

LEONARDO da Vinci [1443 or 1452-1519]

* Many architectural drawings and ms notes are conveniently illustrated and summarized by Luigi Firpo in his *Leonardo, architetto e urbanista*, Turin 1963; but see also Carlo Pedretti, *Leonardo architetto*, Milan 1978; *Leonardo da Vinci: frammenti sull'architettura* ed. Corrado Maltese, in *Scritti rinascimentali di architettura*, Milan 1978; P. C. Marani, *L'architettura fortificata negli studi di Leonardo da Vinci*, Florence 1984; and *Léonard de Vinci ingénieur et architecte*, Montreal 1987

see also under Luca PACIOLI

LIBERATI, Giovanni Antonio

La Caprarola descritta in versi latini Ronciglione 1614

LIGORIO, Pirro [traditionally 1513/14-1583, but more likely 1530-1586]

Libro delle antichità di Roma Venice 1553 [associated with a critical bird's-eye-view map issued simultaneously, with frequent later issues on its own - see A. P. Frutaz, *Le piante di Roma*, Rome 1962, vol 2 plates 25-32, vol 3 pl 672-84]

Thermae Diocletianae et Maximianae Rome 1558

* **Ristretto della villa d'Adriano** [ms descr. and plan] in Antonio del Re, *Dell'antichità tiburtine*, Rome 1611; as **Ichnographia villae Tiburtinae Hadriani Caesaris, olim a Pyrrho Ligorio delineata & descripta postea a Francesco Continio recognita** [text in Latin and Italian] Rome 1751, 1761

* Illustrated ms books on antiquities, incl. architecture, at Paris [Bibl. Nat.], Oxford [Bodleian], Turin [Arch. di Stato] and Naples [Bibl. Naz.] - some drawings from the latter in Erna Mandowsky and Charles Mitchell, *Pirro Ligorio's Roman antiquities*, London 1963; those in Turin as *Libro dei disegni* ed. Caterina Volpi, with notes by Maurizio Calvesi, Rome 1994

LIPSIUS, Justus [Joest Lips - 1547-1606]

De amphiteatro liber Antwerp 1584 [same setting also with Leyden imprint], 1585; Leyden 1589; Antwerp 1598, 1604, 1621; in Graevius, *Thesaurus antiq. roman.*, Leyden 1694-99; and in the many editions of Lips's *Opera*

LOCATELLI, Vincenzo [c1520-c1580]

Invito generale del reparare, fortificare, edificar luoghi Bologna 1575

LOMAZZO, Giovanni Paolo [1538-1600]

Trattato dell'arte de la pittura Milan 1584 [two issues: Pontio, Pontio for Pietro Tini; facs. Hildesheim 1968]; as **Trattato dell'arte della pittura, scoltura, et architettura** 1584 [some copies with an additional chapter after leaf 328], 1585; Rome 1844 [3 vols]; ed. Roberto Paolo Ciardi, first of his two-volume ed. of Lomazzo's *Scritti sulle arte*, Florence 1973

[English trsl. of books I-V by Richard Haydocke] **A tract containing the artes of curious paintinge carvinge & buildinge** Oxford 1598 [facs. Farnborough 1970]

[French trsl. of book I by Hilaire Pader] **Traicté de la proportion** Toulouse 1649

LONGHI [or LUNGHI], Martino [1602-1660]

Epilogismo di architettura Bracciano 1625

LORENZO DE SAN NICOLAS, Fray [Augustinian; 1595-1679]

Arte y uso de architectura [Madrid?] 1633, ?1639, 1667

Segunda parte del arte y uso [Madrid?] 1663 and/or 1664/65 [these may be the same ed.]

[complete work, in 2 vols] Madrid 1736, 1796

[facs. of part 1 of ?1639 and part 2 of 1664/65, with intro by J. J. Martín Gonzalez] Valencia 1989

LORINI, Buonaiuto [1547-c1611]

Delle fortificationi libri cinque Venice 1596 [issued in some 15 copies 'personalized' for presentation to 'Christian princes'], 1597

[German trsl. by David Wormbser] **Fünff Bücher von Vestung Bawen** Frankfurt 1607, ?1621

Le fortificationi ... con l'aggiunta del sesto libro Venice 1609

[German trsl. of book 6] **Das sechste Buch von der Fortification** Oppenheim 1616

[German trsl. of the complete work] **Sechs Bücher ...** Oppenheim 1620

LUPICINI, Antonio [c1530-1598]

Architettura militare Florence 1582; Turin 1585; with Lanteri & Zanchi, Venice 1601

Discorsi militari ... sopra l'espugnazione Florence 1587; with Lanteri & Zanchi, Venice 1601

MACHIAVELLI, Niccolò [1469-1527]

* **Relazione d'una visita fatta per fortificare Firenze** [ms of 1526], first published in vol 2 pp 414-20 of his **Opere** Florence 1782-83; critical ed. by Sergio Bertelli, Milan 1961

MAGGI, Girolamo [c1523-1572] & Giacomo **CASTRIOTTO**

Della fortificatione delle città Venice 1564, 1583/4; selected passages with notes by Paola Barocchi in *Scritti d'arte del Cinquecento* vol 3, Milan-Naples 1977, pp 3468-3505
[the individual contributions of the authors are clearly distinguished]

? [German trsl. by J. Ph. Eboli] **Von der Befestigung der Städte** Giessen 1620 - existence denied by Herzog August Bibliothek, Wolfenbüttel; perhaps confusion with Ebel's German trsl. of Dieterich [q.v.] ?

MAGGIERI, Silvio [active mid-17th century]

Difesa ... ad alcune obbiettioni fatte alla fortificatione italiana Rome 1637

MAGIRUS, Johannes

Compendium fortificatorium oder Kurtzer Begriff der gantzen Fortification Berlin 1600, 1646

MALVICINO-FONTANA, Erasmo [Marquis; active late 16th century]

* Memoranda on fortification and principles of defence [ms. in private collection] described and partly published in J. B. Bury, *An unpublished codex on fortification* in L'architettura militare veneta del Cinquecento, ed. Daniela Lamberini, Milan 1988

MARCANOVA, Giovanni [1410/15-1467]

* **Disegni di Roma antica** [sketchbook of which fragments survive in the Bibl. Estense, Modena, in the Princeton Univ. Library (Garrett ms) and in the Bibl. Nat., Paris] partly published in Christian Huelsen, *La Roma antica di Ciriaco d'Ancona*, Rome 1907; H. van Mater Dennis, *The Garrett MS* in Memoirs of the American Academy in Rome 6, 1927; and E. B. Lawrence, *The Illustrations of the Garrett and Modena MSS*, also in M.A.A.R. 6, 1927

MARCHI, Francesco de' [1504-1576]

Della architettura militare libri tre Brescia 1597 [proofs? without title-page; dedication to Vinc. Gonzaga by Gaspar Lolli refers to 'Novam hanc ce[n]tum sexaginta ... delineationem']; **Della architettura militare libri quattro** 1599, that printing re-issued, each time with altered title-page, in 1600, 1603 and 1609; re-drawn and rearranged, with intro and extensive commentary, by Luigi Marini, 3 quarto text vols and 2 large-folio atlases, Rome 1810 [a few sets, by re-imposition of the text setting, entirely in large folio], reissued 1815 [with simple addition of a letter V after the original date MDCCCX]

MARCUCCI, Jacomo [Crulli da]

Grandezze della città di Roma antiche & moderne Rome 1625, 1628

MARIANO da Firenze, Fra [Franciscan - active late 15th and early 16th century]

* **Itinerarium urbis Romae** [written c1518, ms in Franciscan Archives, Ognissanti, Florence] ed. Enrico Bulletti, Rome 1931 [first guide to contemporary Rome]

MARLIANI, Bartolomeo [died 1560]

Antiquae Romae topographia Rome 1534; as **Topographia antiquae Romae** [edited by Rabelais, with his letter to Joachim Du Bellay, and his index] Lyons 1534 - both octavo; enlarged and illustrated, in folio, as **Urbis Romae topographia** Rome 1544 [two issues, later one deleting dedic. to François I], 1549; Basel 1550; with different illus. and the 1534 text, Venice 1588

together with Pomponius Leto, Publius Victor and/or others in various compilations: *De antiquitatibus urbis Romae libellus*, Basel 1538; *Antiquitatum varii auctores*, Lyons 1552, 1560; J. G. Graevius, *Thesaurus antiq. roman.*, Utrecht-Leyden 1696, etc.; or appended to editions of Livy, Frankfurt 1568, 1578; Paris 1552, 1573

[Italian trsl. of the 1534 version, by Ercole Barbarasa] **L'antichità di Roma** Rome 1548, 1622

[English paraphrase of 1534 version, by Philemon Holland] in Livy, *The Romane historie*, London 1600, 1659

MAROLOIS, Samuel [1572-1627]

Fortification ou architecture militaire The Hague 1614/15; reveüe augmentée et corrigée par Albert Girard, Amsterdam 1627, and as part 5 of the Opera Mathematica [q.v.] but also sold separately without an Opera part-title; also, as part of the Oeuvres mathématiques [q.v.], corr. par Franç. van Schoten [Frans van Schooten], Amsterdam, Blaeu 1628

[Dutch trsl. of the Girard version, by W. D.] **Fortificatie, dat is sterckte bouwing** Amsterdam 1627, 1628 [both using the French engr. title with printed cancel, which is sometimes confusingly missing] and as part of the Opera Mathematica

[German trsl. of the Girard version] **Fortification. Das ist Vestung-Baw** Amsterdam 1627, 1638

[Latin trsl. of the Girard version] **Artis muniendi ...** Amsterdam 1633, 1644 [the latter also issued as part 5 of the 1662 Mathematicum opus]

[English trsl. of the Girard version] **The art of fortification** Amsterdam 1638

Opera mathematica ou oeuvres mathematiques, traictans de geometrie, perspective, architecture, et fortification ... ausquels sont aioints les fondements ... de I: Vred[e]m[an] Vriese, The Hague 1614/15; Amsterdam 1617; rev., augm. et corr. par Albert Girard, Amsterdam 1628, 1638, 1662

Œuvres mathématiques ed. Th. Verbeeck & Franç. van Schoten, Amsterdam, Blaeu 1628

[Dutch trsl. of the Girard version] Amsterdam 1628, 1630, 1651

[Latin trsl. of the Girard version] **Mathematicum opus absolutissimum** Amsterdam 1633, 1649, 1662 [with parts dated 1644 and 1647]

[all Amsterdam editions of the Girard versions were published by Jansonius]

see also the note under Johan VREDEMAN de Vries

MARTINI, Francesco di Giorgio - see FRANCESCO DI GIORGIO Martini

MATTIOLI, Pietro Andrea [1501-1577]

Il magno palazzo del cardinale di Trento Venice 1539; Geneva 1889 [description in verse of the architecture and decoration of palace built 1528-36]

MAUCLERC, Julien [1513-1577]

Le premier livre d'architecture La Rochelle 1600 [author's portrait says 'premiere planche des oeuvres de architecture de Julien Mauclerc ... 1566']; as **Traitté de l'architecture suivant Vitruve** Paris 1648

[English trsl. by Robert Pricke] **A new treatise of architecture** London 1669

? **Méthode nouvelle ... de fortifier les places** Amsterdam 1615

[it is not certain this work is indeed by Julien Mauclerc- only Luigi Marini, *Biblioteca istorico-critica di fortificazione permanente*, lists it, and it is unlikely that it would have escaped further notice if indeed it existed]

MAZZOLARI, Ilario - see José de SIGÜENZA

MELLONI, Antonio [c1500-1549]

Particelli e fragmenti - in Belluzzi 1598, q.v.

MERCATI, Michele [1541-1593]

Degli obelischi di Roma Rome 1584, 1589 [facs., ed. Gianfranco Cantelli, Bologna 1981]

MEYER, Daniel [1576-1630]

Architectura. Vonn Ausstheylung der fünff Seülen Frankfurt 1612 [not to be confused with his here irrelevant *Architectura oder Verzeichnuss allerhand Einfassungen*, Frankfurt 1609, also issued with French title]

MEYNIER, Honorat de [c1570-1638]

Les nouvelles inventions de fortifier les places Paris 1626

[German trsl.] **Fortifications-Baw** Frankfurt 1642

[MIRABILIA ROMAE]

It is impossible to assess what [often very slight] relevance the many Indulgentiae, '*In isto opusculo dicitur*'s, '*Murus civitatis habet*'s, *Mirabiliae*, '*Cose miravigliose*'s [etc. - in their Latin originals or their many translations] might have for this list, without examining each in detail. We therefore refer readers to the most extensive listing of this material so far, which incorporates the material in Schudt's *Guide di Roma* of 1930 as well as that in the Olschki *Choix* volume on Rome of 1936:

Sergio Rossetti, *Rome: a bibliography from the invention of printing through 1899*. Vol 1: The guide books. Florence, 2000 [Biblioteca di bibliografia italiana 157]

MONTANO, Giovanni Battista [1534-1621]

Libro primo. Scielta d. varii tempietti antichi Rome 1624; re-issued or reprinted 1636, 1638

Diversi ornamenti capriciosi per depositi o altari Rome 1625; re-issued or reprinted 1636, 1638

Tabernacoli diversi Rome 1628; re-issued or reprinted 1636, 1638

Architettura con diversi ornamenti cavati dall'antico Rome ?1624 [very doubtful, even if listed by Cicognara], 1636 [two issues]; re-issued or reprinted 1638

Raccolta de' tempi, et sepolcri disegnati dall'antico Rome 1638

Le cinque libri di architettura Rome 1684, 1691
[collected edition of the preceeding five titles]

MONTEMELLINO, Francesco [active mid 16th century]

Discorso sopra la fortificatione del Borgo di Roma
[this text occurs only in Maggi & Castriotto, q.v. The reputed 1548 edition does not exist, that date refers only to the start of the construction of the new Borgo defences]

MONTJOSIEU, Louis de – see Ludovicus DEMONTIOSUS

MORA, Domenico [1539-c1598]

Tre quesiti in dialogo sopra il fare batterie, fortificare una città, et ordinar battaglie quadrate Venice 1567, 1608
[the *Dialogo sopra il fortificare* is on leaves 25r to 52v and has four woodcuts]

MORIGI, Paolo [S.J.; c1550-c1620]

Il duomo di Milano descritto Milan 1642

MUNTINCK, Hendrik [active early 17th century]

Etliche zierliche und unterscheidliche Termen Groningen 1604; Amsterdam c1650-60

NOè, Fra [died 1568; a Servite, often confused with the Franciscan Noè Bianchi]

Viaggio da Venetia al Santo Sepulchro [and similar titles] Venice 1519, 1538, 1546, 1555, 1563, 1566, 1583, 1587, 1598, c1600, 1640, 1647; Bassano 1680, 1728, 1742; Lucca c1700; Treviso 1800

[although the text of these editions - usually accompanied by woodcuts of varying origin - is attributed to Fra Noè, the early appearances, including that in the anonymous *Viazo da Venesia*, q.v., antedate the Venetian prior's visit to the Holy Land, known to have taken place in 1527]

NOYEN, Sebastiaan van [Sebastiano ab Oya - 1523-1557]

Thermae Diocletiani Imp. quales hodie etiamnum extant Antwerp 1558

ODDI, Matteo [1576/7-1626]

Precetti di architettura militare Milan 1626, 1627

ORTI, Ameto

* **La Caprarola** [description in verse, c1585-89] in Fritz Baumgart, *La Caprarola di Ameto Orti*, in Studi Romani 25, 1935

ORTIZ, Blas [Canon of Toledo cathedral; active first half of 16th century]

Summi templi Toletani graphica descriptio Toledo 1549; reprinted as Appendix 2 in vol 3 of *Toletanorum quotquot extant opera*, Madrid 1793

OSIO, Teodato [died c1650]

Dissertatio probans architecturam et agrimensuram Milan 1629; as **De architecturae et agrimensurae nobilitate** 1639

OYA, Sebastiano ab - see Sebastiaan van NOYEN

PACIOLI, Fra Luca [Franciscan; c1445-c1514]

Divina proportione Venice 1509 [two issues: with title or blank first page] [facs. Urbino 1969; with French trsl. by G. Duchesne and M. Giraud, Paris 1980], 1529 [almost certainly a 'ghost']
[the first part contains Pacioli's treatise, in twenty chapters, on architecture [folios D7r to E9v, numbered 23-33]

[critical editions] ed. Constantin Winterberg, Vienna 1896 [facs. Hildesheim 1974]; ed. F. Riva, Milan 1956; ed. Arnaldo Bruschi in *Scritti rinascimentale di architettura*, Milan 1978, pp 85-144

[Spanish trsl. by Ricardo Resta, with intro by Aldo Mieli] Buenos Aires 1946, 1959, pp 145-183

* [ms of the work, in the Ambrosiana Library, Milan] published with intro by Augusto Marinoni, 2 vols, Milan 1982

* [alphabet, intended for stone-cutters and architects] reproduced in Stanley Morrison, *Fra Luca de Pacioli*, New York 1933

* [ms of 1498 with drawings of the geometric forms by Leonardo da Vinci, in Ambrosiana Lib., Milan] Milan 1956

PALISSY, Bernard [c1510-c1590]

Récepte véritable La Rochelle 1563 [facs., ed. E. Rahir, Paris 1919; critical ed. by Keith Cameron, Geneva 1988; ed. Frank Lestringant and Christian Barrataud, Paris 1996], 1564; together Palissy's *Discours de la nature des eaux*, Paris 1636; in the *Oeuvres* ed. Faujas de St Fond &

M. Gobet, 1777; ed. P. A. Cap, 1844; ed. Anatole France, 1880; ed. Benjamin Fillon & Louis Audiat, Niort 1888

[English trsl. by Aurèle La Rocque, with bibliography, in] **Admirable discourses** Urbana IL 1957

PALLADIO, Andrea [di Pietro della Gondola, detto - 1508 or 1518-1580]

Descrittione de le chiese ... di Roma Rome 1554 [facs. in *Five early guides to Rome and Florence*, intro by Peter Murray, Farnborough 1972; in *Andrea Palladio, scritti sull'architettura*, ed. L. Puppi, Vicenza 1988, pages 37-56]

L'antichità di Roma Rome 1554 [facs. in *Five early guides to Rome and Florence*, Farnborough 1972]; Venice 1554, 1555, 1565 [two issues]; Rome 1558, 1570, 1575, 1576, 1587, 1588 [facs. Rome 1973], 1591, 1595; Perugia 1600; Rome 1609; Viterbo 1617; Rome 1620, 1622, 1625, 1629, 1637, 1650 etc. and in *Andrea Palladio, scritti sull' architettura*, ed. L. Puppi, Vicenza 1988, pages 11-36

[as a separately paginated appendix to *Cose miravigliose*] Rome 1563, 1571, 1575 [twice: Blado and Degl'Angeli]; Venice 1575; Rome 1580, 1585, 1587 [three: Diani, Martinelli, and Gigliotto]; Venice 1587; Rome 1589; [Fra Santi's ed.] 1594; [Flaminio da Colle's ed.] 1595, 1596 [twice: Mutio and Facciotto], 1599, 1600 [twice: Fei and Facciotto]; 1615, 1650, 1724, 1750

[as book V of the Architettura] Venice 1711

[Spanish trsl.] in *Mirabilia Romae*, Rome 1575; in *Las cosas maravillosas* 1589 [twice: Diani and Francino], 1648, 1676

[French trsl. by Pompée de Launay, in his *Merveilles de Rome*] Rome 1608; Arras 1612; Rome 1614; Toul 1616; Rome ?1628, 1637, 1646, 1652, 1668, 1676

[Latin trsl.] in *Antiquitates almae urbis Romae*, Rome 1618

[Latin trsl. parallel with Italian, ed. by Charles Fairfax] Oxford 1709

[English trsl. by Giacomo Leoni, appended to the third ed. of his version of the Quattro Libri] London 1742

* **Les bâtimens et les desseins** ed. O. Bertotti Scamozzi [in Italian and French] 4 vols, Vicenza 1776-83 [folio], 1796-97 [quarto]

* **Fabbriche antiche** ed. Richard, Earl of Burlington, London 1730 [not 1732]; as **Le terme dei Romani** ed. O. Bertotti Scamozzi [in Italian and French] Vicenza 1785

I quattro libri dell'architettura Venice 1570 [two issues, one of them in two volumes: *I due libri dell'architettura* – books 1+2 – and *I due primi libri dell'antichità* – books 3+4; the single-volume edition is the 'common' one, basis of the facs., with separate intro by O. Cabiati, Milan 1945, 1951, 1968, 1976, 1980; with intro. by Erik Forssmann, Hildesheim 1979], 1581, 1601 [facs. of Inigo Jones's copy, with transcript of his notes and intro by B. Allsopp, Newcastle-on-Tyne 1970], 1616; as **L'architettura** 1642; with *L'antichità di Roma* as book V, 1711; ed. by Giacomo Leoni, with French trsl. by Fréart revised by Nicholas Dubois whose English trsl. was based on the French, as L'**architettura di A. Palladio** London 1715-20; with French trsl. by 'N.N.' - i.e. G. Fossati & Francesco Muttoni - as vols 2, 5-8 of *Architettura di Andrea Palladio* Venice 1740-48 [facs. Trieste 1973-74], 1769, 1800; ed. with engraved copies of the woodcuts, sponsored by Consul Joseph Smith, c1768-80; critical ed. by P. Marini and L. Magagnato, Milan 1980; in *Andrea Palladio, Scritti sull'architettura* ed. L. Puppi, Vicenza 1988

[facs. editions of the woodcuts only] with notes by L. Magagnato & P. Marini, Milan 1980; with intro and notes by Marco Biraghi, Pordenone 1992

[for Palladio's ms drafts, and for his *Aggionta* to the first book [which dates from after 1570] see *Andrea Palladio, scritti sull' architettura*, ed. L. Puppi, Vicenza 1988

[Italian abstract of book I as] **Delli cinque ordini di architettura** Venice 1746; as **I cinque ordini ... esposti** 1784; Bassano 1803

[Latin trsl. of book I as] **Paladii liber de architectura** Bordeaux 1580

[Spanish trsl. of book I by Francisco de Praves] **Libro primero de la architectura** Valladolid 1625

[French trsl. of part of book I by Pierre le Muet] **Traicté des cinq ordres d'architecture** Paris 1645, 1647, 1764; Amsterdam 1682

[Dutch trsl. of book I by Cornelis Danckerts, after Le Muet] **Verhandeling van de vijf orderen** Amsterdam 1640

[English trsl. of book I by Godfrey Richards] **The first book** London 1663, 1668, 1676, 1683, 1693, enlarged 'With the new Model of the cathedral of St. Paules in London' 1700 [two issues], 1708, 1716, 1721, 1724, 1729, 1733

[English trsl. of book I by Nicholas Dubois, revised by Colen Campbell] **Andrea Palladio's first book** London 1728; as **Andrea Palladio's five orders** 1729

[English trsl. of book I by Isaac Ware] **The first book** [London] 1742

[Italian ed. of book I, as] **Studio elementare degli ordini di architettura** Milan 1818; Leghorn 1820

[German trsl. of books I and II by Georg Andreas Böckler] **Die Baumeisterin Pallas** Nuremberg 1698

[French trsl. by Fréart] **Les quatre livres** [with woodcuts from the original 1570 blocks] Paris 1650 [facs., intro. F. Hébert-Stevens, Paris 1980]; revised by Nicholas Du Bois, as **L'architecture de A. Palladio** [with original text and English trsl. by Dubois, ed. by Giacomo Leoni] London 1715-20; on its own, The Hague 1726

[English trsl. by Nicholas Dubois] **The architecture of A. Palladio** [with original text and French trsl. by Fréart revised by Dubois] London 1715-20; on its own, 1721; with notes from Inigo Jones' copy, and additions, 1742

[French trsl. by 'N.N.' - i.e. G. Fossati & Francesco Muttoni - with original text] vols 2, 5-8 of **Architettura di Andrea Palladio** Venice 1740-48 [facs. Trieste 1973-74], 1769, 1800

[English trsl. by Edward Hoppus] **Andrea Palladio's architecture** London 1732-34 [in parts; then issued with a general title dated 1735], 1736

[English trsl. by Isaac Ware] **The four books** London 1737-38 [facs., intro A. K. Placzek, New York 1965], 1753-55; book I separately, as **The first book** [London] 1742

[Spanish trsl. by J. Fr. Ortiz y Sanz – books 1 and 2 only despite being called] **Los quatro libros** Madrid 1797

[French trsl. by Nicolas Chapuy and Amedée Beugnot, with illus. by Bertotti-Scamozzi, in] **Oeuvres complètes d'André Palladio** Paris 1825-42

[Swedish trsl. by Ebba Atterbom & Anna Mohr Branzell] **Fyra Böcker om Architekturen** Stockholm 1928

[French version of the facs. ed. of 1945, with intro by Cabiati] **Les quatre livres d'architecture** Paris 1960

[German trsl. by Andreas Beyer & Ulrich Schütte] **Die vier Bücher zur Architektur** Zurich 1988

[English trsl. by Robert Tavernor and Richard Schofield] **The four books** Cambridge MA 1997 [with a bibliography including the only mention we have found of the Bordeaux 1580 Latin trsl. as well as some little-known and mostly recent translations from eastern Europe on p xxiii]

PANCIROLI, Ottavio [active late 16th and early 17th century]

I tesori nascosti dell'alma città di Roma Rome 1600; much enlarged and corrected, 1625

PASINO, Aurelio di [active second half of 16th century]

Discorsi sopra il architettura militare Antwerp 1570

[French trsl.] **Discours sur plusieurs poincts de l'architecture de guerre** Antwerp 1579

PELLEGRINI [TIBALDI], Pellegrino [1527-1596]

* **Trattato ... sull'architettura** [mss written between 1586 and 1596, in Paris, Bibl. Nat., Ital. 474, and Milan, Bibl. Ambrosiana, P.246 sup], critical ed. by Giorgio Panizza with intro and notes by Adele Buratti Mazzotta, Milan 1990

* **Architettura di Leon Battista Alberti nel commento di Pellegrini Tibaldi**, critical ed. by Sandro Orlando with intro by Giorgio Simoncini, Rome 1988

PERRET, Jacques [active late 16th and early 17th century]

Des fortifications et artifices Paris 1601 [facs. Unterscheidheim 1971], 1620; as **Architectura et perspectiva des fortifications** Frankfurt 1602 [the frequently cited '1594' edition does not exist, but is the result of confusion: the 1601 title-engraving includes the view of a 1594 siege]

[German trsl.] **Architectura et perspectiva etlicher Festungen** Frankfurt 1602, 1621; Oppenheim 1613

PERUZZI, Baldassare [1481-1536],

* **Trattato di architettura militare** [near-contemporary ms copy of a lost original attributed to Peruzzi, written 1530-36, Accad. di Belle Arti, Florence, Coll. E.2.1.28] with intro and notes by Alessandro Parronchi, Florence 1982

* **Taccuino dei viaggi** [of c1518-19] published in Alfonso Bartoli, *Monumenti antichi di Roma nei disegni degli Uffizi*, vol 2, Rome 1916, pp 195-223

* **Taccuino S IV 7, detto di Baldassare Peruzzi** [ms in Bibl. Communale, Siena] Sovicille 1981

* **Architekturzeichnungen** ed. Heinrich Wurm, Tübingen 1984

PERUZZI, Giovanni Sallustio [died after 1576 but before 1587]

* **Monumenti antichi di Roma:** drawings published in Alfonso Bartoli, *Monumenti antichi di Roma nei disegni degli Uffizi*, vol 4, Rome 1919, plates 372-98

PHILANDRIER, Guillaume [1505-1565]

In decem libros M. Vitruvii Pollionis de architectura annotationes Rome 1544; Paris 1545; Venice 1557; in Ryff & Messerschmidt's ed. of Vitruvius, Strasburg 1550; in his own editions of Vitruvius, Lyons 1552, Lyons/Geneva 1586; in Ryff's editions of Vitruvius, Basel 1575, 1582, 1614; in F. Lemerle, *Architecture et humanisme au milieu du XVIème siècle: Les* Annotationes *de Guillaume Philandrier ... livres I-V*, Tours 1991

PIGAFETTA, Filippo [1533/34-1604]

* **Descrittione de Porti et Fortezze del Regno d'Inghilterra** [ms of 1588 in the Bibl. Nac., Madrid; published in] C. Malfatti, *Cuatro documentos italianos en materia de la expedición de la Armada Invencible*, Barcelona, 1973, pages 11-20

PITTONI, Giambattista

Praecipua aliquot romanae antiquitatis ruinarum monimenta designata Venice 1561, 1575; Rome 1581

POLDO d'Albenas, Jean [1512-1563]

Discours historial de ... Nismes Lyons 1559, 1560 [facs. Marseilles 1976]

POLLAIUOLO, Simone del [known as Il Cronaca - 1457-1508]

* **Libro degli edifici Romani** [part of a sketchbook] published in Alfonso Bartoli, *Monumenti antichi di Roma nei disegni degli Uffizi*, vol 1, Rome 1914, pp 11-20

POMPONIUS Laetus, Julius [1425-1498]

De Romanae urbis vetustate Rome 1510, 1515; in his *Opera* or *Opuscula* Strasburg 1510, 1515; Paris 1511; Mainz 1521

together with Publius Victor, Marliani and/or others in various compilations: *De urbe Roma scribentes*, Bologna 1520; *De Roma prisca et nova varii auctores*, Rome 1523; *De antiquitatibus urbis Romae libellus* Basel 1538; *Antiquitatum varii auctores* Lyons 1552, 1560; and in Livy editions, Frankfurt 1568, 1578; Paris 1573

[Italian trsl.] **L'antiquità di Roma** Venice 1550

PORTA, Giovanni Battista della [1535-1615]

De munitione libri tres Naples 1608

PORTIGIANI, Girolamo [died 1592?]

Prospettiva di fortificationi n.pl., n.d. [title, 16 illus. on 15 pl.]; n.pl., n.d. [title, portrait, 20 illus. on 19 pl.]; Bologna, n.d. [title, portrait, 18 pl.]; Rome 1648 [title, portrait, 20 pl.]

[it seems likely that this very rare suite of fortification designs should always consist of title, portrait, 20 illus. on 19 plates (no. 7 having 2 illus.); that the first two references are to the same edition; and that the first and third refer to incomplete copies]

PORTENARI, Angelo

Della felicità di Padova Padua 1623

POSSEVINO, Antonio [1533-1611]

De architectura tractatus occupies chapters 16 to 18 of book 3 of his *Bibliotheca selecta*, Rome 1593; Venice 1603; Cologne 1607

POTIER d'Estain, Michael

Théorie et pratique des forteresses Cologne 1601

[German trsl.] **Theoria et praxis fortificatorium** Cologne 1602

PUCCINI, Bernardo [1521-1575]

* **Trattato di fortificazione** [MS Magl. XIX 18bis, Bibl. Naz., Florence; dating between 1564 and 1575] transcribed by Daniela Lamberini in *Il principe difeso: vita e opere di Bernardo Puccini*, Florence 1990, pp 273-304 – with notes on Galileo's use of the Trattato on pp 136-38

RABELAIS, François [between 1483 and 1494-1553]

Gargantua Lyons?1534, 1535 - and countless later editions. Chapters 53 and 55, on the Abbaye des Thelemites, are reprinted, with commentary, in Anthony Blunt, *Philibert de l'Orme*, London 1958, pp 8-14. The most careful architectural 'reconstruction' of the Abbaye can be found in Charles Lenormant, *Rabelais et l'architecture de la Renaissance*, Paris 1840. Another, by Léon Dupray, appeared in Arthur Heulhard, *Rabelais, ses voyages en Italie*, Paris 1891, pp 8 and 16

RABICO, Iolante - see Giacomo LANTERI

RADI, Bernardino [1581-1643]

Vari disegni de architettura ornati de porte Rome 1619

RAFFAELLO Sanzio [1483-1520]

* Report of 1519 to Pope Leo X, written by Baldassare Castiglione under Raphael's instructions, outlining the latter's programme for a pictorial reconstruction of ancient Rome. There were three manuscripts of this, slightly different from eachother:

1. [Ms, now lost, formerly in the library of Scipione Maffei] first published in the *Opere volgari e latini del conte Baldessar Castiglione*, Padua 1733, pages 429-436

2. [Codex ital. 37b, Bayerische Staatsbibl. Munich] first published in J. D. Passavant, *Rafael von Urbino*, Leipzig 1858. See *Scritti d'arte del Cinquecento* ed. Paola Barocchi, vol 3, Milan-Naples 1977, pp 2971-85; *Scritti rinascimentali di architettura* ed. Renato Bonelli, Milan 1978, pp 459-84; and Carlo Vecce in *Giorn. Stor. della Lit. Ital.* 173:564, 1996, pp 533-43

3. [Ms in the library of the Castiglioni family of Mantua] first published, in part, by V. Cian in *Arch. Stor. lombardo* 69, 1942, vol 9 pages 70 ff.

[English trsl.] in Carlo Pedretti, *A chronology of Leonardo da Vinci's architectural studies after 1500*, Geneva 1962, pp 162-71

REVESI BRUTI, Ottavio [1570/755-c1642]

Archisesto per formar con facilità li cinque ordini d'architettura Vicenza 1627 [variant issues]

[English trsl. by Thomas Malie] **A new and accurate method** London 1737

RHUMEL, Joh. Pharamundus

Compendium fortificatorium [in German] Nuremberg 1632, 1644

RICCI, Fray Juan [Benedictine; 1600-1681]

* **Tratado breve de perspectiva y arquitectura** [probably written between 1629 and 1641] published by Elias Tormo y Monzo and Enrique Lafuente Ferrari in *La vida y la obra de Fray Juan Ricci*, 2 vols, Madrid 1930

RIDINGER, Georg [1568-after 1628]

Architectur des Maintzischen Churfürstl. Neuen Schlossbaues St. Johannspurg zu Aschaffenburg Mainz 1616; annotated ed. by Hans-Bernd Spiess, Aschaffenburg 1991

RINALDINI, Giovanni [1557-1620]

Conclusione che non si devono fare le piazze basse ne' fianchi delle fortezze Messina 1610

RITRATTI di Roma antica [moderna] - see Pompilio TOTTI

RIVAN, Antoine

L'art de fortifier les places Paris 1628, ?1634, ?1636

RIVIUS, G. H. – see Walther Hermann RYFF

ROJAS, Cristóbal de [1555-1614]

Teorica y practica de fortificacion Madrid 1598

Compendio ... de fortificacion Madrid 1613

* **Sumario de la milicia antigua y moderna** [ms R34728, Bibl. Nac., Madrid; completed at Cadiz, 20 Jan. 1607; includes rules for fortification in part 2] published, with facs. of the other two works, as **Tres tratados sobre fortificación y milicia**, with intro by Ramón Gutiérrez, and a second vol. reprinting Eduardo Mariátegui's 1880 biography of Rojas, Madrid 1985

RORICZER, Mathes [c1440-1492/95]

puechle[in] d fiale[n] gerechtikait [Regensburg] 1486 [facs. Regensburg 1923; with intro by Ferdinand Geldner, Wiesbaden 1965]; as **Das Reissbüchlein der Massbretter** von Matthias Roritzer, in Karl von Heideloff, *Die Bauhütte des Mittelalters in Deutschland*, 1844, pp 101-16; ed. with trsl. into modern German & intro by A. Reichensperger, Trier 1845

[Italian trsl. by Franco Borsi, in] *Per una storia della teoria delle proporzioni*, Florence 1967, pp 150-230

[With English trsl. and intro in] Lon R. Shelby, *Gothic design techniques*, Carbondale IL 1977

there seems also to be a 4-page tract on gables, reacting to that of Schmuttermayer [q.v.], c1590, but we have not been able to locate a copy or even find precise details

ROVERE, Francesco Maria I della [fourth Duke of Urbino, 1490-1538]

Discorsi militari Ferrara 1583

* **I discorsi sopra le fortificazioni di Venezia** [ms in Bibl. Marciana, Venice, written c1536-8] ed. Elisa Viani, Mantua 1902

RUBENS, Peter Paul [1577-1640]

Palazzi moderni di Genova [Antwerp 1622] [facs. Unterscheidheim 1969]

Palazzi antichi di Genova Antwerp c1626

[both works combined] Antwerp 1652 [facs., with appendix, Novara 1955], 1663 [facs., intro by A. A. Tait, New York 1968], 1708; as **Architecture italienne ... 3e éd.** Amsterdam 1755. Facs., ed. Hildebrand Gurlitt, Berlin 1924; Genoa 1955; reduced, with intro and notes by H. Schomann, Harenberg 1982. Critical ed. by M. Labó, Genoa 1970. See also *Rubens e Genova*, exhib. catalogue, Palazzo Ducale, Genua 1977

RUIZ, Hernán [c1505-1569]

* **El libro de arquitectura de Hernán Ruiz el joven** [ms of c1560] intro by Pedro Navascués Palacio, Madrid 1974

see also VITRUVIUS: Spanish trsl. of book I, c1560

RUSCONI, Giovanantonio [c1520-1587]

Della architettura libri dieci [illustrations for a projected edition of Vitruvius, begun c1553; published with captions by the publisher] Venice 1590 [facs. Farnborough 1968; with intro by A. Bedon, Como 1996]; as **I dieci libri d'architettura** 1660 [two issues]

RYFF, Walther Hermann [c1500-c1548]

Der fürnembsten, notwendigsten, der gantzen Architectur angehörigen ... Künst, Nuremberg 1547 [facs., intro by E. Forssmann, Hildesheim 1981], as **Der Architecture fürnembsten ...** 1558; as **Bawkunst** Basle 1582

see also VITRUVIUS: German trsl. 1, 1548

SAGREDO, Diego de [Chaplain to Queen Juana; died 1527/28]

Medidas del Romano [text probably completed 1521] Toledo 1526 [facs., ed. F. Z. Lucas and E. P. de Leon, Madrid 1946, in 200 copies; ed. by Santiago de Sebastián, Cali (Colombia) 1967; with intro by L. Cervera Vera, Valencia 1976; Churubusco (Mexico) 1977]; Lisbon 1541 [three printings; facs. Lisbon 1915, in 100 copies not for sale], Jan. 1542, June 1542; Toledo 1549 [facs. with intro by F. Marías and A. Bustamante, Madrid 1986], 1564

[French trsl., with additions which were retained in all later editions in both French and Spanish] **Raison d'architecture antique, extraicte de Vitruve & autres anciens architecteurs** Paris [1531/37], 1539, 1542, 1550, 1555 [two issues]; as **De l'architecture antique** 1608

SALVIATI, Giuseppe PORTA, detto [c1520-c1575]

Regola di far perfettamente col compasso la voluta et del capitello ionico et d'ogn'altra sorte Venice 1552; reprinted in Gio. Ant. Selva, *Delle differente maniere di descrivere la voluta ionica*, Padua 1814

[Latin trsl.] in Poleni & Stratico's ed. of Vitruvius, vol 1, Udine 1825, pp 269-73

[Russian trsl. by V. P. Zubov] Moscow 1938

SAMBIN, Hugues [1515/20-c1601]

Oeuvre de la diversite des termes Lyons 1572 [two issues]

SANGALLO, Antonio da [the elder - 1453 or 1455-1534]

* **Codex Strozzi** published with intro in Alfonso Bartoli, *Monumenti antichi di Roma nei disegni degli Uffizi*, vol 1, Rome 1914, plates 66-76

SANGALLO, Antonio da [the younger - 1484-1546]

* Preface to a projected trsl. of Vitruvius [Codex Magliabechianus C, Bibl. Naz., Florence] in A. Gotti, *Vita di Michelangelo*, Florence 1876, vol 2 pp 179 ff; in G. Giovannoni, *Antonio da Sangallo il Giovane*, Rome 1959, vol 1, pp 395-97; critical ed. by Paola Barocchi in *Scritti d'arte del Cinquecento* vol 3, Milan-Naples 1977, pp 3028-31

* **The architectural drawings of Antonio da Sangallo the younger and his circle**, ed. Christoph L. Frommel & Nicholas Adams, vol 1 [fortification, machines, festival architecture], New York & Cambridge MA 1994

SANGALLO, Giuliano Giamberti da [1443 or 1445-1516]

* **Il libro di Giuliano da Sangallo,** codice Vaticano Barberiniano Latino 4424, ed. Christian Huelsen, 2 vols, Leipzig & Turin 1910; Vatican City 1984

* **Il tacuino senese di Giuliano da Sangallo** ed. Rodolpho Falb, Siena 1902

* **Giuliano da Sangallo: i disegni** ed. Stefano Borsi, Rome 1985 [a small-size edition of the two preceeding items]

SANSOVINO, Francesco [1521-1583]

Dialogo di tutte le cose notabili che sono in Venetia Venice 1556, 1558; as **Delle cose notabili che sono in Venetia** 1562, 1564 [and perhaps later]

Venetia citta nobillissima ... Venice 1581, 1604; continued by Giustiniano Martinioni, 1663 [facs. Venice 1968; with intro by J. Fletcher, Farnborough 1972]

SARAYNA, Torello [died 1550]

De origine et amplitudine civitatis Veronae Verona 1540 [illustrations by Caroto, q.v.], 1560 [unillustrated]

[Italian trsl. by Orlando Pescetti, unillustrated] Verona 1586 [facs. Bologna 1975]

SARDI, Pietro [c1560 - after 1639]

Corona imperiale dell'architettura militare Venice 1618; Bologna 1689

[German trsl.] **Corona imperialis** Frankfurt & Hamburg 1622/26; Frankfurt 1623, 1640, 1644

[German trsl. of book I by Joh. Ludwig Gottfried] **Corona imperialis** Frankfurt 1626

[French trsl.] **Couronne imperiale** Frankfurt 1622/3

Discorso per il quale ... si rifutano tutte le fortezze ... fatte con semplice terra Venice 1627

Corno dogale dell'architettura militare Venice 1639

[the *Corno Dogale* is not, as often alleged, a second edition of the *Corona Imperiale* – nor does a 1638 edition actually exist, despite its appearance in several library catalogues where the date of the dedication has been used instead of that of the colophon, there being no date on the title]

Il capo de' bombardieri Venice 1641 [both separately and as part of an anthology called *Fucina di Marte*]
[pages 84-146 examine fortification in detail, from the artillerist's point of view]

Discorso sopra la necessità e utilità dell'architettura militare Venice 1642

SARTI, Antonio

? **L'aurora delle opere di fortificazione** Venice 1626

La reale et regolare fortificatione Venice 1630, ?1634

SARTI, Paolo

La simmetria dell'ottima fortificazione Venice 1630

SATTLER, Heinrich Johann

Von Vestungen, Schantzen und gegen-Schantzen Basel 1619; as **Fortificatio** 1627

SAVORGNANO, Mario [Count of Belgrado; c1513-1574 or 1597]

Arte militare terrestre, e maritima Venice 1599, 1614
[only pages 234-240 deal with fortification]

SAVOT, Louis [c1579-1640]

L'architecture françoise des bastimens particuliers Paris 1624, 1642; with notes and additions by Fr. Blondel, 1673, 1685 [facs. Geneva 1973]

SCALA, Giovanni [active from 1588]

Delle fortificationi Rome 1596 [two issues], 1627, 1642

SCAMOZZI, Giovan Domenico [c1526-1582]

Discorso intorno alle parti dell'architettura [in Serlio, *Tutte le opere*] Venice 1584, 1600 [facs. Oviedo 1986], 1618/19 [facs. Ridgewood NJ 1964; Sala Bolognese 1970]

[more recently the *Discorso* has been cogently attributed to Vincenzo Scamozzi, despite Giovan Domenico's name being printed in the heading. See inter alia L. Puppi, *Scrittori Vicentini d'architettura del sec. XVI*, Vicenza 1973, pp 97-105]

[Spanish trsl. by Fausto Diaz Padilla] in *Todas las obras de arquitectura y perspectiva de Sebastián Serlio* [trsl. from the Venice 1600 ed.] Oviedo 1986, pp 213-17

SCAMOZZI, Vincenzo [1548 (not 1552)-1616]

Discorsi sopra l'antichità di Roma [written to accompany 40 engravings by Giovanni Battista Pittoni, q.v., first published Venice 1561, and in part copied after Hier. Cock and thus perhaps going back to Maerten van Heemskerck, q.v.] Venice 1581, 1582 [facs., intro L. Olivato, Milan 1991], 1583 [two issues]

L'idea della architettura universale Venice 1615 [facs. Ridgewood NJ 1964; Sala Bolognese 1982]; Piazzola 1687; Venice 1694, 1714; ed. S. Ticozzi & L. Masieri, Milan 1811, 1838

[the *Idea* contains only books 1-3 and 6-8 of the intended ten books - fragments of the other books and their illustrations were published by Danckerts at Amsterdam in 1661: see below. Of the six published books, only two continued to attract wide interest: book 3, on houses and palaces, i.e. civil architecture, and especially book 6, on the orders. Book 2 includes a treatise on military architecture]

[Dutch trsl. of book 6 by Cornelis Danckerts] **Grontregulen der Bow-konst** Amsterdam 1640

[Dutch trsl. of book 3 by Corn. Danckerts, with the original woodcuts, not yet incl. those for the unpublished books 4 and 5: the 28 plates correspond solely to those of book 3 in the original 1615 ed.] **Het voorbeelt der algemeene bowkonst** Amsterdam 1658

[Dutch trsl. by Corn. & Dancker Danckerts] **Bouwkundige werken,** begrepen in 8 boeken, Amsterdam 1661 [this ed. includes fragments of the hitherto unpublished books 4 and 5, together with some of their illus.; books 4-6 are dated 1661; book 3 is in the ed. of 1658; others not dated]

[abridged Dutch trsl. of book 6 by Simon Bosboom, with his own scale for calculating measurements] **De grondt-regulen der bouw-kunst ... Cort onderwijs van de vijf colommen** Amsterdam 1657/8, 1670; ed. Dirk Bosboom, 1682, 1686, 1694, c1715, c1720, 1760; ed. Caspar Philips, 1774, 1784, 1816, 1821, 1854

[summary Dutch trsl. of book 6 after Danckerts' 1661 ed., by Joachim Schuym; plates only, with separately printed captions and Schuym's own scale] **De grondt-regulen der Bouw-konst** Amsterdam 1661, 1662 [two issues], 1677, 1694

[English trsl. after Schuym's summary version, by W. F(isher)] **The mirror of architecture** London 1669, 1671, 1676, 1687, 1693, 1700, 1708, 1721, 1734, 1752

[English trsl. after Bosboom, by Robert Pricke] **A brief and plain description of the five orders** London 1676

[German trsl. of book 6 after Schuym's summary version] **Die Grund-Regelen der Bau-Kunst** Amsterdam 1664, 1665

[German trsl. of book 6 after Danckerts' 1640 ed.] **Grund-Regeln der Bau-Kunst** Sulzbach/Nuremberg 1678, 1697

[German trsl. of books 3 and 6 after Danckerts, plates re-engraved by Wilhelm Pfann] **Klärliche Beschreibung Der fünff Säulen-Ordnungen** Sulzbach/Nuremberg 1677/8, 1697

[abridged French trsl. of book 6, by Aug. Ch. d'Aviler] **Les cinq ordres d'architecture** Paris [folio] 1685; [octavo] 1710, 1730, 1764

[French trsl. of book 6 by d'Aviler, and of the rest of Danckerts' 1661 ed. by Samuel du Ry] **Oeuvres d'architecture** Leyden 1713; The Hague 1736

[French compendium by Jombert] **Oeuvres d'architecture** Paris 1764

[compendium by Bald. Orsini] 3 vols, Perugia 1803; with plate volume added, Milan 1838

* **Tacuino di viaggio da Parigi a Venezia** [written 14 March to 11 May 1600] ed. Franco Barbieri, Venice 1959

SCHADAEUS, Oskar [1586-1626]

Summum Argentoratensium templum ... Beschreibung dess ... Münsters zu Strassburg Strasburg 1617

SCHILLE, Hans van [c1521-1586]

Form und weis zu bauwen Antwerp 1573, 1578 [both anonymous], 1580, c1593

SCHMUTTERMAYER, Hans [died after 1518]

Fialenbüchlein [Nuremberg, 1489 - only one incomplete copy known]; ed. A. Essenwein in Anz. Knd. Dt. Vorzeit NS28, 1881, cols 65-78

[With English trsl. & intro in] Lon R. Shelby, *Gothic design techniques*, Carbondale IL 1977

SCHRÖTER, Hans

Extrakt in der Fortification Celle 1633

SCHULTZ, Georg

Fortification und Mess Kunst Erfurt 1639

SCRIVá, Pedro Luis [Comendador, Order of Malta; c1485-c1540]

* **Apologia en excusacion y favor de [sus] fábricas** [illus. ms of c1538] ed. Eduardo Mariátegui, Madrid 1878

SERLIO, Sebastiano [1475 (or possibly after 1490)-1553/54]

Hoping that reference will be facilitated thereby, the various books have been listed in order of their number instead of in date order of their appearance. For each book, the subsequent separate editions and translations follow. Collected editions appear at the end. Unless listed otherwise, all editions are in folio

books I-II

Il primo libro d'architettura. Le premier livre d'architecture [bilingual, the French trsl. by Jean Martin] Paris 1545, sheets reissued in two variants [with or without the 1545 colophon] 1590

[in Italian only] **Il primo libro ...** Venice c1551, 1560

[it is doubtful that the first set of Sessa editions, which consists of these undated books and issues of 3, 4, 5 all dated 1551, were ever meant to be sold as anything other than a 'complete works' - they are never found separately in contemporary bindings

The 1560 edition forms part of the second set of Sessa issues, which presents more of a problem since its four volumes have widely separate dates - 1560, 1562, undated, and 1559 - for which there is no reasonable explanation other than that they did indeed appear at different times and must have been available separately - the fact that the signatures through volume 3-5 follow a definite sequence may just have been copied from the earlier set. Even so, the existence of two copies of volume 3 on its own - in Paris and Cambridge libraries - is hardly conclusive evidence for separate sale, as both are in 19th century bindings, and although entries in early inventories [e.g. those of Jorge Manuel Theotocopoulos of 1621 and G. B. Paggi of 1627] record the existence of single books, these entries are not sufficiently explicit to resolve the issue. Even the purchase by Vincenzo Borghini of a copy of book II [perspective] on its own from the Giunti of Florence, 14 Nov 1552, can not be said to settle the matter]

[Flemish trsl. by Pieter Coecke van Aelst] **Den eersten** [tweeden] **boeck van architecturen** Antwerp 1553, 1558
[in this edition the two books do indeed have separate title-pages and imprint]

[English trsl. taken from the 1611 collected edition] book I as **A new-naturalized work of a learned stranger** London 1657; book II as **A book of perspective and geometry** London 1657

* * *

book III

Il terzo libro ... le antiqutà di Roma Venice 1540, title corrected to **le antiquità** 1544, 1551 [see note to *books I-II*, above], 1562 [see continuation of same note]

[Flemish trsl. by Pieter Coecke van Aelst] **Die aldervermaerste antique edificien** Antwerp 1546

[French trsl. by Pieter Coecke van Aelst] **Des antiquités** Antwerp 1550

* * *

book IV

Regole generali di architettura Venice 1537, 1540, 1544, 1551 [see note to *books I-II*, above], c1562 [see continuation of same note]

[Flemish trsl. by Pieter Coecke van Aelst] **Generale reglen der architecturen** Antwerp 1539 [facs., ed. Rudi Rolf, Amsterdam 1978]; as **Reglen van metselrijen** 1549

[German trsl. by Jacob Reichlinger] **Die gemaynen Reglen von der Architectur** Antwerp 1542, 1558

[French trsl. by Pieter Coecke van Aelst] **Reigles generales de l'Architecture** Antwerp 1542, 1545, 1550

* * *

book V

Quinto libro ... diverse forme de tempii sacri [bilingual, the French trsl. by Jean Martin] Paris 1547

[in Italian only] **Quinto libro ...** Venice 1551 [see note to *books I-II*, above], 1559 [see continuation of same note]

[Flemish trsl. by Pieter Coecke van Aelst] **Den vijfsten boeck ... diversche formen der templen** Antwerp 1553, 1558

* * *

* *book VI*

Habitationi di tutti li gradi degli homini [ms of c1541-47 in Avery Library, New York, AA.520.Se.619.F] facs., ed. Myra Nan Rosenfeld, intro James Ackerman, as **Sebastiano Serlio on domestic architecture**, New York 1978

[ms of c1547-54 in Bayerische Staatsbibliothek, Munich, Codex icon. 189] facs., ed. Marco Rosci and A. M. Brizio, Milan 1966, with separate vol. of commentary by Anna Maria Brizio; facs. with Spanish trsl. by Fausto Diaz Padilla, Oviedo 1986

* * *

book VII

Architecturae liber septimus ... Il settimo libro ... di molti accidenti, che possono occorrer' al architetto [bilingual] Frankfurt 1575; Venice 1663

* [ms of c1541-50 in Oesterreichische Nationalbibliothek, Vienna] partly published in Myra Nan Rosenfeld, *Serlio's drawings in Vienna for his seventh book*, in Art Bulletin 56, 1974, and in T. Carunchio, *Dal VII libro di Serlio: XXIIII case per edificar nella villa*, in Quaderni dell'Ist. di Storia dell' architettura, fasc. 127-32, 1976

*'book VIII'

Castrametatione [ms of c1546-54 in Bayerische Staatsbibliothek, Munich, Codex icon. 190] partly published in W. B. Dinsmoor, *The literary remains of Serlio*, in Art Bulletin 24, 1942; in P. Marconi, *L'VIII libro inedito di Serlio*, in Controspazio 1:4/5, 1969; and in June Gwendolyn Johnson, *Sebastiano Serlio's treatise on military architecture*, UCLA thesis, Los Angeles 1984; facs. with Spanish trsl. by Fausto Diaz Padilla, Oviedo 1986

* * *

'Extra'

Livre extraordinaire ... auquel sont demonstrees trente portes rustiques ... et vingt autres d'oeuvre delicate [engravings, with long captions] Lyons 1551, 1560[/61?]

[in Italian & French] Lyons 1551

[in Italian only] **Extraordinario libro** Lyons 1551, 1558, 1560; [with plates reversed] Venice 1557, 1558, ?1559, 1560, 1561, 1566, 1567, 1568; [reduced to quarto, with woodcuts] 1566, ?1572, 1584, 1600, 1618, 1619

* * *

books III + IV

Venice 1619 [in National Union Catalogue, but likely to be just two parts of the 1619 Opere

NB: Marcolini published the first edition of book III in March 1940 and the second edition of book IV in February 1540, and republished both in 1544. The intention seems to have been to see the two books as a set, as were books I and II subsequently. This is corroborated by several surviving examples of Marcolini sets of III and IV bound together – see John Bury, *Bibliographical notes* in Acts of Serlio Conference [Vicenza 1987], Milan 1989, p 92.

[Spanish trsl. by Francisco de Villalpando] **Tercero y quarto libro de architectura** Toledo 1552 [facs. with intro by George Kubler, Valencia 1977], 1563, 1573

[the two books have separate sets of signatures and colophons and book IV has a title-page of its own and could theoretically have been sold separately, but again there is no evidence that this ever happened. The colophon borders repeat those of the Marcolini editions, indicating a derivation from those rather than the then more recent Sessa editions]

* * *

books I-V

[Sessa issued two uniform editions, the second much better edited than the first, which are usually referrred to as 'collected works' but were never titled thus - see the notes on these issues under *books I-II*, above] Venice 1551 [but no date on books I/II], revised 1559-62 [no date on book IV]

[Flemish trsl. by Pieter Coecke van Aelst] **Den eersten** [... vijfsten] **boeck** Amsterdam 1606, reissued 1616, ?1626, ?1636

[German trsl. from Flemish, by Ludwig Koenig] **Von der Architectur fünff Bücher** Basel [variant issues dated] 1608 and/or 1609; Frankfurt 1672-74 [gives author's name as the anagram Liserus - therefore sometimes unwittingly referred to as a 'plagiarism']

[English trsl. 'out of Dutch'] **The first** [... fift] **booke of architecture** London 1611 [facs. with intro. by A. E. Santaniello, New York 1970, 1980; another, with bibliography, 1982]

[English trsl. by Vaughan Hart and Peter Hicks] **Sebastiano Serlio on architecture** vol I [books I-V] New Haven CT 1996

* * *

books I-V + Extra

Libro primo [... estraordinario] **d'architettura** [quarto] Venice 1566, 1572

[Latin trsl. by G. C. Saraceni] **De architectura libri quinque** Venice 1569/68 [two issues: with or without a 16-page preface]

[in Latin and Italian, the Liber Estraordinario with only 11 plates] **Architettura ... in sei libri divisa** Venice 1663

* * *

book I-V + Extra + VII

Tutte le opere [quarto] Venice 1584 [facs. with an essay on post-WW2 Serlian criticism by F. Irace, Sala Bolognese 1970], 1600 [facs. with intro by Carlos Sombricio, Oviedo 1986], 1618/19 [facs. of copy with notes by John Webb, Ridgewood NJ 1964]

[Spanish trsl. by Fausto Diaz Padilla] **Todas las obras de arquitectura y perspectiva de Sebastián Serlio de Bolonia** with intro by Carlos Sambricio, Oviedo 1986

* * *

books VI-'VIII'

[transcriptions of mss, with facs. of illustrations] intro and notes by Francesco Paolo Fiore [the Munich mss of VI and VIII] and Tancredi Carunchio [the Vienna ms of VII] Milan 1994

* * *

books VI-'VIII' and Extra

[English trsl. by Vaughan Hart and Peter Hicks] **Sebastiano Serlio on architecture** vol 2, New Haven CT [forthcoming]

* * *

SHUTE, John [died 1563]

The first and chief groundes of architecture London 1563 [facs. London 1912, with intro by Lawrence Weaver; Farnborough 1964], 1579/80, 1584, 1587 [the latter using sheets of the previous edition]

SIGÜENZA, José de [Jeronymite, librarian of the Escorial; c1544-1606]

La historia de la Orden de San Geronimo 3 vols Madrid 1595-1605 [Sigüenza's description of the Escorial is in vol 3]; 2 vols Madrid 1907-9 [the description in vol. 2]. On its own, the description appeared as **Fundación del monasterio de El Escorial** Madrid 1927; with illus. and intro by F. C. Sainz de Robles, Madrid 1963; with illus. and intro by Antonio Fernández Alba, Madrid 1988.

[Italian trsl., with a few omissions, by Ilario Mazzolari] **La reali grandezze dell'Escuriale di Spagna** Bologna 1648, 1650
[Padre Mazzolari, of whom nothing is known (it may be an assumed name), presents himself as author instead of translator, suppressing the name of Sigüenza. No library catalogue, not even those of the British Library and the Library of Congress, has recognized the pretence, making Mazzolari's plagiarism one of the most successful ever perpetrated]

SOLIS, Virgil [1514-1562] - see Jacques ANDROUET DU CERCEAU

SOLMS, Reinhard, Graf zu [1491-1562]

Eyn gesprech ... welcher masse ein vester bawe fürzunemen Mainz 1534/35; as **Ein kurtzer Ausszug** Cologne 1556

SPECKLE, Daniel [1536-1589]

Architectura von Vestungen Strasburg 1589, revised and augmented 1599 [facs. Portland OR 1972], 1608; Dresden 1705, 1710, 1712, 1736

SPINI, Gherardo [born 1538]

* **I tre primi libri sopra l'instituzioni de' Greci et Latini architettori** [ms of c1567-68] published with intro and notes by Cristina Acidini in *Il disegno interrotto*, Florence 1980, pp 11-201

STEVIN, Simon [1548-1620]

Sterckten-bouwingh Leyden 1594; Amsterdam 1624; in *Les oeuvres mathématiques* ed. Albert Girard, Leyden 1634

[German trsl. by Gothard Arthus] **Festung-Bawung** Frankfurt 1608, 1623

T

TARTAGLIA, Nicolò FONTANA, detto [c1500-1557]

Quesiti, et inventioni diverse [book 6, *Del modo di fortificar le città*, ff 69r-75v] Venice 1546; 'con una gionta al sesto libro nella quale si demostra un primo modo di redurre una città inespugnabile' 1554 [facs., ed. Arnaldo Masotti, Brescia 1959], ?1562 [very dubious: all copies listed under this date start with leaf 5 and end on leaf 94, they have no imprint and are bound up with editions of the Noua Scientia of 1558 or 1583 - the latter combination resulting in mentions of a 1583 Quesiti, which is equally unproven]; in the **Opere** 1606

[German trsl. of book 6 by Walther Ryff] in his **Architectur** 1547 etc., q.v.

[French trsl. of book 6, with its 'gionta'] **Livre VI des Demandes et Inventions** Reims 1556

[German trsl. of book 6, with its 'gionta', by A. Böhm] in *Magazin für Ingenieure und Artilleristen*, vol 4, 1778, pp 11-64

TENSINI, Francesco [1581-1630]

La fortificatione Venice 1623/24; sheets reissued with altered title-page 1630; with altered title-page and re-set last page including colophon 1655

THETI [Theti or Tetti], Carlo [1529-1589]

Discorsi di fortificationi [unauthorized] Rome 1569; revised and greatly enlarged authorized version as **Discorsi delle fortificationi** [books 1 and 2] Venice 1575; Rome 1585; books 3-6 published as **Discorsi ... divisi in libri quattro** Venice 1588; books 1-2 reprinted, books 7-8 added, and the 1588 edition of books 3-6 inserted between them, as **Discorsi ... divisi in libri otto** 1589; [complete] Vicenza 1617

[French trsl.] **Discours sur le faict des fortifications** Lyons 1589

THIRY, Léonard – see Jacques ANDROUET DU CERCEAU

THORPE, John [c1565-c1655]

* **The book of architecture in Sir John Soane's Museum** [plans and elevations of houses, executed between c1596 and 1623] ed. John Summerson, Glasgow 1966

TIBALDI - see Pellegrino PELLEGRINI [TIBALDI]

TOLOMEI, Claudio [Archbishop of Curzola, Bishop of Toulon; 1492-1555]

De le lettere lib. sette Venice 1547 - and 12 or more editions over the next 50 years, as well as a French trsl., Paris 1572; in 4 vols, Fermo 1781-83.

See in particular the letters of 14 Nov. 1542 to Agostino Landi, of 20 June 1544 to Gabriel Cesano and of c1546 to Francesco Sansovino - the Landi and Cesano letters in Bottari & Ticozzi, *Raccolta di lettere*, vol 2 pp 1-17 and vol 5 pp 107-27, Milan 1822 - undated letters to Antonio

Rusconi and Francisco Sansovino in vol 5 pp 97-100 and 138-40; Landi [in Latin trsl.] in Poleni & Stratico's ed. of Vitruvius, vol 1, Udine 1825, pp 197-201; Landi, Cesano and Sansovino in Paola Barocchi, *Scritti d'arte del Cinquecento*, vol 3, Milan-Naples 1977, pp 3037-46, 3123-33 and 3047-48 resp.

TOTTI, Pompilio [editor - born c1590]

Ritratto di Roma antica Rome 1627, 1633, 1645, 1654, 1688, text revised 1689; as **Descrizione ...** 1697, 1707

Ritratto di Roma moderna Rome 1638, 1645, 1652, 1689; as **Descrizione ...** 1697, 1707

[both works together, as volume 1 and 2] Rome 1708, 1719, 1727, 1739, [in 3 volumes] 1745, 1750, 1765

[Dutch trsl. of both] **Afbeelding van 't oude [nieu] Romen** Amsterdam 1661

[German trsl. of both] **Abgebildetes altes [neues] Rom** Arnhem 1662

TRISSINO, Giangiorgio [1478-1550]

* **Dell'architettura** [ms treatise in Bibl. Braidense, Milan] in L. Puppi, *Un letterato*, in Arte Veneta 25, 1971; with notes in Paola Barocchi, *Scritti d'arte del Cinquecento* vol 3, Milan-Naples 1977, pp 3032-36; critical ed. by Giacomo Semenzato in Elena Bassi et al., *Pietro Cataneo, Giacomo Barozzi da Vignola, Trattati*, Milan 1988, pp 19-29; a fragment had been printed from an 18th century transcript in the *Nozze* book of the Peserico-Bertolini marriage, Vicenza 1878. See also L. Puppi, *Scrittori Vicentini d'architettura del sec. XVI*, Vicenza 1973, pp 79-86

UGONI[O], Pompeo [died 1614]

Historia delle stationi di Roma Rome 1588

VALLE, Giambattista della [1470-?1550]

Vallo Naples 1521; Venice 1524, 1528, 1529, 1531, 1535, 1538/9, 1543, 1550, 1558, 1564

[French trsl.] Lyons 1529, ?1531, ?1534, 1554

[Greek verse trsl. by Leonardis Phortios, with the original Venice woodcuts] **Poièma neon** [*graece*] Venice 1531 [only two copies known]

[Spanish version by Diego de Alaba y Viamont] as part one of his **El perfeto capitan** Madrid 1590

VANDELVIRA, Alonso de [c1545-1625]

* **Tratado de arquitectura** [written 1575/91, ms in Madrid School of Architecture] ed. Geneviève Barbé-Coquelin de Lisle, 2 vols, Albacete 1977

VASARI, Giorgio [the elder - 1511-1574]

Le vite Florence 1550 [2 vols]; enlarged and illustrated with woodcut portraits, 3 vols - basis of most later editions, 1568; also in **Le opere**, notably Rome 1759-60 [7 vols, ed. Bottari, portraits engr. by Bartolozzi; reprinted] Leghorn [vol 1] 1767 and [vols 2-7] 1771-72; ed. Guglielmo della Valle [12 vols] Siena 1791-93; ed. Gaetano Milanese, Florence 1878-85 [9 vols; reprinted] 1906 [facs. 1973]

* **Il libro de' disegni** [Vasari's collection of master drawings, including numerous architectural designs by or attributed to Brunelleschi, Francesco di Giorgio Martini, Andrea Sansovino, Bramante, Giuliano Antonio and Francesco Sangallo, Peruzzi, Scamozzi and others; disbound, in the Uffizi, Florence] published as *Il libro de' disegni del Vasari*, ed. Licia Ragghanti Collobi, 2 vols, Florence 1974. See also her *Vasari: Libro de' disegni -- architettura* in Critica d'arte, fasc 127, Florence 1973

VASARI, Giorgio [the younger - 1562-1625]

* **La città ideale di Giorgio Vasari il giovane** ed. by Virginia Stefanelli with intro by Franco Borsi, Rome 1970 [includes a *Libro di diversi piante*, architectural drawings for an ideal city, dated 1598 - Uffizi Gab. dei disegni mss 4529-4594 - and *Piante di chiese [palazzi e ville] di Toscana e d'Italia*, undated - Uffizi Gab. dei disegni mss 4715-4944]

* **Porte e finestre di Firenze e di Roma** [late 16c, Uffizi Gab.dei disegni mss 4595-4714] published with intro and notes by Franco Borsi in *Il disegno interrotto*, Florence 1980, pp 293-321

VIGNOLA, Giacomo Barozzi da [1507-1573]

Regola delli cinque ordini d'architettura [Rome 1562]

It seems superfluous to give a list here of the many editions, translations and adaptations of Vignola, since Maria Walcher Casotti has already done this comprehensively for 514 existing editions, laying to rest a number of non-existent 'ghosts' from older lists. Her bibliography appears on pages 527-575 in *Trattati di architettura* vol 5 part 2, Milan 1985

We can add only one minor item to her list:

[Portuguese trsl. by J. C. Sequeira, with notes] **Breve tractado das cinco ordens** Lisbon 1841

and one other [apart from her own edition in *Pietro Cataneo, Giacomo Barozzi da Vignola, Trattati*, Milan 1985] which appeared after her deadline:

[Spanish trsl. of 1593] facs. with intro by A. R. Gutierrez de Ceballos, Valencia 1985

VILLALPANDO, Juan Bautista [S.J.; 1552-1608]

Templi Hierosolymitani commentarius et imaginibus illustratus, as vol 2 [1604] of Hier. Prado & Io. Bapt. Villalpando, *In Ezechielem explanationes*, Rome 1596-1604
[these are the published dates, but the colophon on p 655 of part 2 vol 2 says 'Anno Domini MDCV. à qua die tres simul tomi incipient evulgari']

[Spanish trsl. of vol 2 by José Luis Oliver Domingo, with glossary, illus. and indices] **El templo de Salomón segun Juan Bautista Villalpando**, with two further volumes: a facs. of Prado's 1593 ms *Compendio* about his own reconstruction of the temple [ms BMR108, Houghton Lib., Harvard] as **El templo de Salomón segun Jerónimo de Prado**, and a volume of studies by André Corboz, J. A. Ramirez and others, as **Dio arquitecto**, accompanied by a disk containing indices of words and themes in Villalpando's work; 2 vols folio, 1 large 4to, and disk, Madrid 1991

VILLE, Antoine de [1596-1656]

Les fortifications Lyons 1628, 1629, 1640; Paris 1636, 1666, 1696; as **La fortification** Amsterdam 1672, 1675, 1696

[German trsl.] **Die Festungs-Bau-Kunst** Amsterdam 1676; as **Vollkommener Ingenieur** Frankfurt 1760

Descriptio portus et urbis Polae Venice 1633
[illustrates the Roman remains and his own work on the fortifications]

VIOLA Zanini, Gioseffe [1575/80-1631]

Della architettura libri due Padua 1629, [with appendix] 1677-78
[vol 2, with the title **Della nuova simmetria degli cinque ordini d'architettura**, can also occur separately]

VITRUVIUS Pollio, Marcus [c90-c20 b.C.]

[The titles of most editions as well as translations are simple variants on 'the ten books of architecture', in various languages: to transcribe all these would not have helped the legibility of the listing]

editions

* **De architectura libri decem** [ed. Giovanni Sulpicio] Rome [between 1486 and 16 August 1487 at which date the copy now in Corpus Christi, Oxford, was bought in Rome. Both Eucharius Silber and Georg Herolt have been named as the printer]

[ed. unknown] Venice, De Pensis, November 1495 [colophon on f. 64r] or Florence, Ant. Francisci, 1496 [colophon on f. 70v]

[ed. unknown, although Fra Giocondo, Vittor Pisano and Giorgio Valla have all been suggested] Venice 1497

[ed. Fra Giovanni Giocondo] Venice 1511 [folio; first illustrated version - after which nearly all editions are illustrated], adapted to octavo size Florence 1513, 1522

[ed. Scipio de Gabiano ?] Lyons 1523 [often described as a pirated version of Giocondo's 1522 edition, but it is not]

[ed. Walther Ryff ? and Georg Messerschmidt] Strasburg 1543; with Philander's notes added, 1550

[ed. Guillaume Philandrier, with first Life of Vitruvius] Lyons 1552 [facs. of books 1-4, with French trsl. and commentary by Frédérique Lemerle, Paris 1999]; Lyons/Geneva 1586

[ed. Daniele Barbaro] Venice 1567

[ed. Johannes de Laet, together with a number of related texts] Amsterdam 1649

[ed. B. Galiani, with his Italian trsl.] Naples 1758

[ed. August Rode] Berlin 1800, plate volume 1801; without the 66-page dictionary, Strasburg 1807

[ed. Jo. Gottlieb Schneider] 3 vols Leipzig 1807-8; single-vol Tauchnitz ed. 1836; with Galiani's Italian trsl., Venice 1854 [ed. Giovanni Poleni and Simone Stratico, together with a number of early Vitruviana] 4 vols, often in 8, Udine 1825-30

[ed. Ch.-L. Maufras, with his French trsl.] 2 vols, Paris 1826, 1847, 1850-53

[ed. Aloisio Marini, simultaneous with his Italian trsl.] 4 vols Rome 1836

[ed. Nisard, with Perrault's trsl, and with other texts] **Celse, Vitruve, Censorin ...** Paris 1852, 1857, 1877

[ed. Valentin Rose & Hermann Müller-Strübing, after Schneider] Leipzig [Teubner] 1867; [ed. Rose alone] 1899

[ed. Otto Holtze] Leipzig 1869, 1892

[ed. Auguste Choisy, with his French trsl.] 4 vols, often in 2, Paris 1909 [facs. with intro by Fernand Poullon 1971]

[ed. Fr. Krohn] Leipzig [Teubner] 1912

[ed. Frank Granger, with his English trsl.] Cambridge [Loeb Library] 2 vols 1931-34, 1955-56, 1962, 1970; ed. I. D. Rowland, New York 1998

[ed. Ugo Fleres, with his Italian trsl.] 2 vols, Milan 1933

[ed. of books 1-7 by Silvio Ferri, with his Italian trsl.] Rome 1960

[ed. Curt Fensterbusch, with his German trsl.] Darmstadt 1964, 1981, 1991

[ed. in progress, with French trsl.] book 1 by Philippe Fleury, Paris 1990; book 3 by Pierre Gros, Paris 1990; book 4 by Pierre Gros, Paris 1992; book 7 by B.Liou, M. Zuinghedau & Marie Therèse Cam, Paris 1995; book 8 by Louis Callebat, Paris 1973; book 9 by Jean Soubiran, Paris 1969; book 10 by Louis Callebat & Pierre Fleury, Paris 1986

[ed. Pierre Gros, with Italian trsl. by Antonio Corso and Elisa Romano] 2 vols. Turin 1997

* * *

* for details on extant **manuscripts** of the text see Carol Herselle Krinsky, *Seventy-eight Vitruvius manuscripts*, in Jnl of the Warburg & Courtauld Institutes 30, 1967 [no longer a complete list: Ms Krinsky listed three more later, and we know of three manuscripts in private hands, also one at Chatsworth and two in the Vatican - Pal. lat 867 and Ottob. lat. 1234]

* many manuscript editions still unpublished include the one by Joannes Franciscus Fortyna, architect of Padua, dedicated to Cosimo de' Medici [ms Med. Palat. 51], and the Italian versions made and illustrated by G. B. de Sangallo [mss Bibl. Corsiniana 1846 and 2093]

* * *

* **paraphrases** and translations into Italian of passages from books 1-6, together with copies of pre-Carolingian, possibly late antique, illustrations, inserted by Buonaccorso Ghiberti into his *Zibaldone*, have been given a critical ed. in Gustina Scaglia, *A translation of Vitruvius*, in Trans. Amer. Philos. Soc. 69, Feb. 1979

* * *

* **epitomes** of Vitruvius were published in

  Raffaello Maffei Volaterrano, *Commentariorum rerum urbanorum libri*, Rome 1506; Basel 1544; Frankfurt 1603; etc.

  Cassiodorus, *De quattuor mathematicis*, Paris 1540

  Gaudenzio Merula, *Memorabilium*, 1556

  Antonio Possevino, *Bibliotheca selecta*, Rome 1593; Venice 1603; Cologne 1607; etc.

* * *

translations

* [Italian trsl. 1, by Marco Fabio Calvo, c1514-15, 'ad instantia di Raffaello', in the margins of a copy of the 1511 ed. - ms Bayerische Staatsbibl. Munich, Codice Italiano 37] in *Vitruvio e Raffaello* ed. Vincenzo Fontana and Paolo Morachiello, Rome 1975

[Italian trsl. 2, by Cesare Cesariano, with books 9 and 10 finished by Bono Mauro and Benedetto Jovio] Como 1521 [facs. New York 1968; reduced-size facs. with intro by Carol Herselle Krinsky, Munich 1969; with intro by A. Bruschi, Adriano Carugo and Paolo Fiore, Milan 1982]; ed. F. Lutio Durantino [often wrongly credited as translator], in the lay-out and with the illustrations of the 1511 edition, Venice 1524, 1535

* [Italian trsl. 2, by Cesare Cesariano: ms 9/2790 Sección de Cortes della Real Acad. de la Historia, Madrid, ed. I. B. Agosti] **Volgarizzamente dei libri IX [capitoli 7 e 8] e X di Vitruvio** Pisa 1996

[Italian trsl. 3, books 1-5 only, by G. B. Caporali] **Architettura con il suo comento et figure** Perugia 1536

[French trsl. 1, by Jean Martin] **Architecture, ou art de bien bastir** Paris 1547 [facs. Ridgewood NJ 1964], 1572; with different illustrations, Geneva 1618

[German trsl. 1, by Walther Ryff] **Vitruvius teutsch** Nuremberg 1548 [facs., intro Erik Forssmann, Hildesheim 1973; [with Philander's notes added] Basel 1575, 1582, 1614

[Italian trsl. 4, by Daniele Barbaro] folio, illustrated by Palladio, Venice 1556; reworked, quarto, 1567 [facs. with intro by Manfredo Tafuri, and analysis of the differences between Barbaro's 1556 and 1567 editions by Manuela Morresi, Milan 1987], 1584, 1629, [sheets reissued] 1641, 1854

[French abbrev. trsl. 1, by Jean Gardet and Dominique Bertin] **Epitome, ou extrait abrégé** Toulouse 1559; Paris 1559, 1565, 1597

* [Spanish trsl. of book 1 on pp 55-76 of Hernán Ruiz] **El libro de arquitectura** [ms of c1560] ed. Pedro Navascués Palacio, Madrid 1974

[Spanish trsl. 1, by Miguel de Urrea] Alcala 1582 [facs., intro. by Luis Moya, Valencia 1978]

[French trsl. 2, by Claude Perrault] Paris 1673, 1684; ed. Tardieu and Cousin, 2 vols, 1837, 1859, 1877; with new illus. by Jean Vital Prost, 1946; ed. André Dalmas, with added photographs, 1965, 1967

[French abbrev. trsl. 2, by Perrault] **Abrégé** Paris 1674, 1768; Amsterdam 1681 [=1691]

[English trsl. of the Perrault digest 1] **An abridgment** London 1692

[English trsl. of the Perrault digest 2, by Abel Boyer; with Moxon's abridged Vignola] **The theory and practice of architecture** London 1703, sheets reissued 1729

[Italian trsl. of the Perrault digest, by Carlo Cataneo, omitting the military architecture] **Compendio** Venice 1711, with Barbaro's commentary, as **L'architettura generale** 1747 [two issues], 1794

[English trsl. by Robert Castell: although sometimes listed as having been printed in 1730, with a reprint in 1747, this projected translation never came further than a prospectus issued in 1728]

[German trsl. of the Perrault digest by M. Müller] **Vitruvii architectura, in das kurze verfasst** Nuremberg 1757

[Italian trsl. 5, by Galiani, with his Latin ed.] Naples 1758; Italian version only, Siena and Naples 1790; reduced version, Milan 1832, 1844; with Schneider's Latin ed. 1854

[Spanish ed. of the Perrault digest by Joseph Castañeda] **Compendio** Madrid 1761 [two issues; facs. with intro by Joaquin Bérchez Crómez, Murcia 1981], 1790

[English trsl. 1, by William Newton] books 1-5 only, London 1771; complete, 2 vols, 1791

[Spanish trsl. 2, by Joseph Ortiz y Sanz] Madrid 1787 [facs. Oviedo 1974]

[Russian trsl. of the Perrault digest, by Vas. Bazenov and F. Karsavin] **Ov arkhitekture** Moscow 1790-97

[German trsl. 2, by August Rode, based on Galiani's ed.] 2 vols, **Des Vitruvius Baukunst** Leipzig 1796

[German trsl. 3, by H. Chr. Genelli] Brunswick 1801

[Italian trsl. 6, by Baldassare Orsini, but also using Galiani's trsl.] 2 vols, Perugia 1802

[English trsl. 2, by William Wilkins] **The civil architecture** London 1812-17]

[French trsl. 3, by De Bioul, based on Galiani's ed.] Brussels 1816

[French trsl. 4, by Ch.-L. Maufras, with his Latin ed.] 2 vols, Paris 1826, 1847, 1850-53

[English trsl. 3, by Joseph Gwilt] London 1826, 1839; with new steel engravings 1860, 1874; essay by Aberdeen added 1880, [1890s]

[Italian trsl. 7, by Carlo Amati] 2 vols, Milan 1829-30

[Italian trsl. 8, by Quirico Viviani] 11 parts, often in 3 vols, Udine 1830-32

[Italian trsl. 9, by Aloisio Marini, simultaneous with his Latin ed.] 5 vols, Rome 1836

[German trsl. 4, by J. E. Hummel] Berlin 1840

[Polish trsl. 1, by Edw. Raczynski, based on Schneider's ed.] 2 vols and atlas, Wroclaw 1840

[German trsl. 5, by C. Lorentzen, books 1-5 only] Gotha 1857

[German trsl. 6, by Franz Reber] Berlin 1865, 1885, 1897

[Hungarian trsl. by Bela Fuchs and Jusztin Bodiss] Budapest 1898

[German trsl. 7, by Jakob Prestel] 4 parts, often in 2 vols, [Heitz] Strasburg 1900/01, 1912-14; Baden-Baden 1959

[German trsl. 8, by L. Sontheimer] Tübingen 1908

[French trsl. 5, by Auguste Choisy, with his Latin ed.] 4 vols, often in 2, Paris 1909, [facs. with intro by Fernand Poullon] 1971

[German trsl. 9, by Adalbert Birnbaum] Vienna 1914

[German trsl. 10, by Heinrich Röttinger] Strasburg 1914

[Dutch trsl. by J. H. A. Mialaret] Maastricht 1914

[English trsl. 4, by Morris Hickey Morgan - first edition illustrated with photographs] Cambridge MA 1914; New York 1960

[German trsl. 11, by W. Sackur] Berlin 1925

[English trsl. 5 by Frank Granger, with his Latin ed.] **On architecture edited from the Harleian manuscript 2767** 2 vols, Cambridge 1931-34, 1955-56, 1962, 1970; ed. I. D. Rowland, New York 1998

[Italian trsl. 10, by Francesco Pellati] Rome 1932

[Italian trsl. 11, by Ugo Fleres, with his Latin ed.] 2 vols, Milan 1933

[Italian trsl. 12, by R. di Lüttichau] Fano 1934

[Russian trsl. by V. P. Zubov, with Barbaro's commentary] Moscow 1938

[German trsl. 12, by E. Stürzenacker] Essen 1938

[Italian trsl. 13, by G. Guenzati - much abbreviated] Milan 1943

[Japanese trsl.] Tokyo 1943

[Italian trsl. 14, by C. J. Moa] Milan 1945

[German trsl. 13, by Herbert Koch] Baden-Baden 1951

[Czech trsl. by Alois Otoupalk] Prague 1953

[Spanish trsl. 3, by Agustín Blánquez] Barcelona 1955, 1970

[Polish trsl. 2, by Kasimir Kumaniecki and Piotr Bieganskyj] Warsaw 1956

[Italian trsl. 15, by Silvio Ferri, with his Latin ed. - books 1-7 only] Rome 1960

[German trsl. 14, by Curt Fensternbusch, with his Latin ed.] Darmstadt 1964, 1981, 1991

[Rumanian trsl. by G. M. Cantacuzino] Bucarest 1964

[French trsl. 6, by André Dalmas] Paris 1965

[Spanish trsl. 4, by Carmen Andreu] Madrid 1973

[Italian trsl. 16, by Fontana and Morachiello] Rome 1975

[Italian trsl. 17, by Giovanni Florian] Pisa 1978

[Czech trsl. by Alois Otoupalik] Prague 1979

[Italian trsl. 18, by G. Scaglia] Florence 1985

[German trsl. 15, by Frank Zöllner] Worms 1987

[Italian trsl. 19, by Gabriele Morolli] Florence 1988

[French trsl. 7, in progress, with Latin ed.] book 1 by Philippe Fleury, Paris 1990; book 3 by Pierre Gros, Paris 1990; book 4 by Pierre Gros, Paris 1992; book 7 by B. Liou. M. Zuinghedau & Marie Therèse Cam, Paris 1995; book 8 by Louis Callebat, Paris 1973; book 9 by Jean Soubiran, Paris 1969; book 10 by Louis Callebat & Pierre Fleury, Paris 1986

[German trsl. 16, by Heiner Knell] Darmstadt 1991

[Italian trsl. 19 by Antonio Corso and Elisa Romano, with the Latin ed. by Pierre Gros] 2 vols. Turin 1997

[English trsl. 6 by Ingrid Drake Rowland, with commentary and illus. by Thomas Noble Howe] Cambridge 1999

[French trsl. 8 by Frédérique Lemerle] book 1-4 only, with facs. of these books in the Lyons 1552 ed., Paris 1999

* * *

VOLPAIA, Bernardo della [active first half of 16th century]

* 'Coner' sketchbook [Soane Museum, London] published as **Sixteenth century drawings of Roman buildings attributed to Andreas Coner**, ed. Thomas Ashby, in Papers of the British School at Rome 2, 1904, and 6, 1913 [facs. New York, 1971]
[The authorship of Volpaia was established by Tilmann Buddensieg in *Bernardo della Volpaia und Giovanni Francesco da Sangallo* in Römisches Jahrbuch für Kunstgeschichte 15, Rome 1975]

VREDEMAN de Vries, Johan [hence, colloquially, both 'Hans' and 'Jan'; 1527-?1606] and his son Paul [1567-1630]

Until recently the bibliography of the many Vredeman de Vries 'books' – often suites of plates with little or no text – and of their complicated [partial] re-use by Hondius and Marolois, has been all but impenetrable. Now, however, we have Peter Fuhring's two volumes in the Hollstein series [vols 47 and 48, Rotterdam 1997] to guide us through: volume 47 covers 1550-1571, volume 48 the period 1572-1630

WAGNER, Bartholomaeus [of Augsburg]

Der Layen Kirchenspiegel Thierhaupten 1593/94, 1596; Constance 1595

WHITEHORNE, Peter [active 1550-1563]

Certain waies for the ordering of Souldiers ... And also Fygures of certaine new plattes for fortification of Townes London 1562 [facs. together with Whitehorne's trsl of Machiavelli's Arte of warre, Amsterdam & New York 1969], 1573, 1588
[leaves 16r-23r are 'of fortification'; 4 of the 7 plates are copied from Zanchi]

WOTTON, Sir Henry [1568-1639]

The elements of architecture London 1624 [facs., intro and notes by F. Hard, Charlottesville VA 1968; other facs. Farnborough 1969; New York 1970], 1903 [privately printed in 350 copies for S.T. Prideaux, from the dedication copy to Charles Prince of Wales (Charles I) in the British Library]; in John Evelyn's trsl. of Fréart's *Parallel*, London 1722, 1723, 1733; abridged as **The ground-rules of architecture** in Scamozzi's *Mirror of architecture*, London 1671, 1676, 1687, 1693, 1700, 1708, 1721, 1734, 1752

[Latin trsl. by Joh. de Laet] as pp 1-30 in De Laet's ed. of Vitruvius, Amsterdam 1649; in Poleni & Stratico's ed. of Vitruvius, vol 1, Udine 1825, pp 205-36

[Latin trsl. in *Reliquiae Wottonianae*] London 1651 [facs. Springfield MA 1897], 1654, 1672, 1685

[Spanish trsl.] Madrid ?1698 [from De Laet's Latin], 1903

[**WYSSENBACH,** Rudolph]

Wunderbarliche kostliche Gemält, ouch eigentliche Contrafacturen mancherley schönen gebeüwen, Zurich 1561, 1566; as **Wahrhaffte Contrafacturen etlich alter und schöner Gebäuen** Zurich 1562; as **Architectura antiqua, Das ist, wahrhaffte und eigentliche Contrafacturen etlich alter schönen Gebeuwen** 1596, 1627 [see for these two the notes under Blum, *V: Columnae*], 1662
[an unnumbered suite of 16 imaginary classical buildings of which some are attributable to Hans Blum, including one dated 1545, while two others bear the monogram of Rudolph Wyssenbach, another that of J. Wyssenbach and the date 1558. Most are reversed copies after Androuet Du Cerceau, with additional decorations by Heinrich Vogtherr]

ZANCHI, Giovanni Battista Bonadio de' [1515-c1586]

Del modo di fortificar le città Venice 1554, 1556, 1560; with Lanteri and Lupicini, 1601; selected passages with notes in Paola Barocchi, *Scritti d'arte del Cinquecento* vol 3, Milan-Naples 1977, pp 3449-67

[French trsl. by François de la Treille] **La manière de fortifier villes** Lyons 1556 [notwithstanding the existence of a copy with 1550 ownership note]

* [English trsl. from the French, by Robert Corneweyle] **The maner of fortification of cities** [ms dated 1559] facs., intro by Martin Biddle, Farnborough 1972

ZUALLART, Jean [Giovanni Zuallardo; 1541-1634]

Il devotissimo viaggio di Gerusalemme Rome 1587 [quarto], 1595 [octavo]
[includes views, plans and details of pilgrimage churches]

[French version] **Le tresdevot voyage de Ierusalem** Antwerp 1604 [dubious], 1608, 1626

[German trsl.] Cologne 1606, 1609, 1659

ZUCCARO, Federico [1542/43-1609]

L'idea de' pittori, scultori et architetti Turin 1607 [facs., ed. D. Heikamp, *Scritti d'arte di Federico Zuccaro*, Florence 1961]; Rome 1768

Lettera a prencipi et signori amatori del dissegno, pittura, scultura et architettura Mantua 1605

ZúCCOLO, Gregorio [active last quarter of 16th and first quarter of 17th century]

I discorsi Venice 1575
[includes a *Discorso intorno alla fortificationi*: text on pages 231-269, 3 woodcuts on pages 270-272]

Printed in the United States
by Baker & Taylor Publisher Services